Super Physics for
Super Technologies

Super Physics for Super Technologies

Replacing Bohr, Heisenberg, Schrödinger & Einstein

Benjamin Thomas Solomon

BSc (Eng., Aston),, MA Operations Research (Lancaster), MBS(Finance, Dublin)

Propulsion Physics, Inc.

2015

Super Physics for Super Technologies
Replacing Bohr, Heisenberg, Schrödinger & Einstein

Propulsion Physics, Inc.
2015

ISBN-13: 978-1508948018
ISBN-10: 1508948011

www.PropulsionPhysics.com

Foreword

It is my pleasure to introduce this work by Benjamin Solomon, and to verify as to the motives of the principle investigator, as a result of a several year interaction with Benjamin.

Benjamin has been re fashioning a conceptual basis, i.e. the standard model of physics especially the classical to quantum correspondence, which up to a point works. But with serious glitches. Here are two of them for the record.

Robert Geroch in "Geometrical Quantum Mechanics", 1974, intertwined mathematics and physics in his notes, with the startling caveat on page 67 of his manuscript, that "Quantum mechanics is a smeared out version of classical mechanics when the amount of smearing is governed by a certain constant, Planck's constant with the additional qualification that the quantum system reduces to the classical system". This usual program works up to a point except when one gets to classical systems such as the Logistics map, which even today does not have an 'obvious' quantum analogue. While attempts to put in quantum chaos do exist, and I know some of the researchers doing them, the entire enterprise has a very forced nature which to this day really does not work.

Benjamin, to his credit has been trying to fashion an alternative, and I have tried to encourage him in this direction, especially in lieu of what I wrote above.

Secondly, is the nature of singularities. Until no less than Hawking admitted that the Black holes are not quite what he thought they were, i.e. as of 2014, and now the theoretical community is adapting to the fall out. But before Hawking, Penrose as of his book, 'The Road to Reality' , in his section 16.3 brought up some implications of the axioms of choice which mean specifically that singularities have to be rethought. And that their nature is not quite what the theoretical community thought it was.

Indeed they do. And Benjamin has been attempting to in his own fashion to redo what is known in GR, so that singularities are not an essential part of cosmology.

Benjamin is the second researcher I have met who has tried to consider a non singular cosmology. The first was Cristi Stoica, whom I met in 2010.

Dr. Stoica met me in Dice 2014, showing me his work in non singular GR, and Benjamin has been attempting, from a different vantage point to do non singular cosmology. I credit both of them for having the insight to make this necessary attempt.

To those who state that there is no flaw in basic cosmology, I say that history has a nasty habit of burying presumed necessary conditions which are irrefutable and not to be changed.

As of the 19th century, largely due to Maxwell, and his E and M theory the aether, i.e. ruled supreme, and that it took Michaelson and Morley and their experiment, which blew this edifice apart, and which led to decades of theoretical anarchy, not resolved in part, to today. That is, I have encountered researchers who have tried to present DARK ENERGY as yet another updated version of the Aether, and have been unimpressed.

Benjamin, with his own resources, and his own will has been trying to fashion together alternatives. That is, as of today, SUSY is not yet experimentally confirmed and may never be, which will lead to yet another massive theoretical crisis. In which then, alternatives to ironclad orthodoxy will have to be found and acted upon.

As an amusing side glance, one of the main work horses of theoretical physics, was in the 1950s derided as not necessary, i.e. one of the main advocates of group methods, was called 'the group pest'. Then irony of ironies, the presumed discovery of quarks commenced, with one very humorous side note. The definition of a group, including multiplicative identity was not done exactly correctly. Even Gell Mann, of Caltech did not quite do it right in the early 1960s.

Today, no one allegedly in their right minds would question 'group theory'

As of 130 years ago, no one dared to question the Aether. That is until experiment ruined it for good.

Happily for the discipline, physics remains an experimental science. In which I say that I appreciate my good friends investigations, since then maybe what we think of as physically sacred ground will have to be changed in a quick moment. And then, much as Sophius Lie of the 19th century, his Lie groups, was viewed as a curiosity, became a work horse, some of what Benjamin is doing will move from curiosity to work horse. With that, I encourage the reader to preview what Benjamin is doing and to keep an eye out for the inevitable dynamics of what an experimentally based science, Physics is really doing.

Andrew Beckwith, PhD
New York

Acknowledgements

I would like to thank Dr. Andrew Beckwith, Professor, Chongqing University, China, for all the encouragement he has given these past many years I've known him. For reviewing the material in this book, and providing additional suggestions, and advice.

Contents

Super Physics for Super Technologies

1. The Origins of the Book

Solomon started his research in late 1999 with an electrical circuit that could change its weight by a small amount. That led him to the review and testing of simple anomalous gravitational effects. Was there a common gravitational thread running through these observations? Could gravity modification as a physics and a technology be discovered?

It was not until 2007 with extensive numerical modeling that Solomon finally discovered $g=\tau c^2$, the massless formula for gravitational acceleration. Solomon showed that $g=\tau c^2$ worked for gravitation, mechanical and electromagnetic accelerations. Something that neither String Theories nor Quantum Theories had been able to achieve. These findings have been published in his 12-year study titled, *An Introduction to Gravity Modification*.

$g=\tau c^2$ was proof that there were another set of rules by which the Universe could be modeled by, as none of our contemporary theories, Relativity, String & Quantum (RSQ) could deliver a theory of gravity that does not require mass. Such a theory would be invaluable to developing the next generation space launch vehicles that don't expel combustion exhaust.

With this book, *Super Physics for Super Technologies*, Solomon asks and answers the question, since $g=\tau c^2$ can we rethink physics? The answer is *yes*. In the 1st chapter Solomon proposes a fourth approach to forces, Spatial Gradients. In the 2nd chapter Solomon proves that gravity is not due to mass but due to matter. This removes the need to model gravity as a function of elementary particle physics of mass, and more importantly facilitates the development of non-particle-based macro level theories for propulsion engine development.

In the 3rd chapter Solomon proposes a new type of matter *Variable Electric Permittivity* (VEP) matter, that holds the nucleus together. In the 4th chapter Solomon replaces Schrödinger wave function, and as a result is able to propose a new electron shell model that fits Quantum Theory's exactly. Thus, in chapter 5 Solomon proposes a new standard model for particles; and in chapter 6 a unified spectrum independent photon behavior.

Solomon's work has led to the inference that it is possible to develop a new technology that will neutralize any current or future stealth technology.

Super Physics for Super Technologies

2. A Universal Approach to Forces

Abstract:- This paper presents a universal approach to forces that is valid for gravity, electromagnetism, electric and magnetic fields; and unlike quantum theory, does not require an exchange of particles.

The change in the center C_F of the spatial gradient of a field is sufficient to determine forces. This shift in C_F, either towards or away from the source of the field is interpreted as attraction or repulsion, respectively.

To evaluate the C_F, requires particles and fields to occupy space. Therefore, it requires a process model that informs as to how the field's spatial gradient is translated with respect to the space it occupies. The mathematical solutions to these process models, inform the velocities and accelerations present in the field's local space.

This approach led to the discovery of the massless equation for gravitational acceleration $g = \tau c^2$, where τ a spatial gradient, is the change in time dilation (as the dimensionless[ix] ratios t_v/t_0 or t_a/t_0) divided by the change in that distance present in the local gravitational field. This new $g = \tau c^2$ relationship has been shown to be correct for electromagnetic and mechanical accelerations, too.

More importantly this process model explains *why* force is orthogonal to both electron velocity and magnetic field lines. Electric field lines are not repulsive in the electromagnetic transverse wave, therefore, assuming that Nature is consistent everywhere, this Spatial Gradient Center of Field approach was applied to the electric field.

Just by summation of the electric field strengths or its spatial gradients, it was shown (within the limits of the numerical models) with respect to the shift in C_F, that exact equations exists that govern velocity and acceleration. And by analogy, for magnetic monopole fields, too.

Thereby making repulsive electric & magnetic field lines falsifiable and lending itself to new particle models by eliminating the need for Poincaré stresses in charged particle models.

2.1 Premise

This paper is the 11th in the 16-year [1-10][i] investigation into the feasibility of determining an alternative mechanism as to how forces are transmitted. This paper proposes that a particle's velocity and acceleration is evidenced by the shift in the Center of Field C_F of the local field's spatial gradient. This field could be gravitational, electromagnetic, electric, magnetic or mechanical motion. It is proposed that the deformation of this field results in the shift in the Center of Field C_F, just as altering the shape of an object would alter its Center of Mass, C_M. The magnitude and direction of the shift in this C_F governs the strength and direction (attraction or repulsion) of the resulting motion of this Field. Using the Center of Mass concept, the Center of Field C_F of a field F that ranges from lower limit L to upper limit U, is defined as

$$C_F = \int_L^U P(x)x\,dx \left/ \int_L^U P(x)\,dx \right.$$

(1)

Where P is the property of the Field used to evaluate the Field's C_F. The relevant field property depends upon the type of deformation applied to this field. If the deformation of the field property is non-linear then the spatial gradient of the Field's property P or dP/dx is the parameter used to estimate the Field's C_F. If the deformation of the field property is linear then the field's property P is the parameter used to evaluate the Field's C_F as the spatial gradient of P is zero.

2.2 Introduction

Steinhardt [11] stated that the Planck Space Telescope data shows that the Universe is simpler than had been thought for three reasons (1) the deviations in the Cosmic Microwave Background is within the limits of quantum theory, (2) don't see gravitation waves i.e. inflation and (3) don't see the effects of string theories. Esftathiou [11] concurred adding that theories leading to infinities reflects a crisis in physics and a resolution to these problems would involve revolutionary physics.

Steinhardt [11] suggested that "progress might not come from where we are looking" and a possible avenue for new research is the long cyclical universe. To add to this debate, using Hubble photographs of gamma ray burst, Nemiroff [12] showed that quantum foam could not exists. One could probably qualify this finding by restating that quantum foam could not exists in the absence of matter.

Solomon [2] proposed contemporary physics can be categorized by three types of particles, inelastic & point-like (quantum theory), tensile (strings) and compressive. Assuming that particles were compressive Solomon [6 & 9] proved that a new equation for gravitational acceleration, $g = \tau c^2$, equation (16), that does not require a prior knowledge of the amount mass. Concurring with Steinhardt & Esftathiou's [11] that both string and quantum theories require revisions.

Therefore, pursuing Steinhardt & Esftathiou's [11] need for a different approach to gravity, one can expand the scope of this need by requiring that new theories on gravity are consistent in the forecasts of how Nature works from the perspective of (1) distant cosmology, (2) near field local gravity probes, and (3) local gravity modification. Broadening the scope to include both, near field local gravity probes and gravity modification, introduces more avenues for exploration. Various researchers [13 - 16] using near field gravity probes have shown that it is very difficult to concur on a specific value for the gravitational constant G. This strongly suggests that gravity does not behave exactly as one would like it to and therefore, there is more to gravity than we currently understand or recognize.

Gravity modification introduces another dimension to the gravity problem. What if gravity was not caused by mass? Podkletnov [17 -18] had reported gravity shielding effects above a spinning superconducting ceramic disc. Podkletnov's results have not been reproduced because other experimenters' [19 - 20] ceramic discs cracked before reaching Podkletnov's disc spin requirements. This is unfortunate as (16) suggest that these experiments ought to be successful. Solomon [4 & 7] had suggested an approach to deconstructing Podkletnov's experiments with the expectation that future experimenters can avoid the mistakes of past experiments [19 - 20]. The importance of these gravity modification experiments is that they can lead to the confirmation that mass is not necessary for gravitational fields and therefore, matter and not mass is the cause of gravitational fields.

Thus, an approach to new avenues of research lies in theories similar to General Relativity in that they are not particle based. In Newtonian gravity, rest mass generates gravitational effects directly by the translation of potential energy to kinetic energy. In General Relativity [21], all sources of both energy and momentum contribute to generating spacetime curvature and that the energy-momentum tensor $T^{\alpha\beta}$ is the source of the spacetime curvature. This curvature then causes the effect of a body falling in a gravitational field.

In this paper, unlike Newtonian gravity or General Relativity, the importance of the shape of spacetime lies in the fact that it informs what time dilation and length contraction are, as these two parameters are the minimum information one requires to determine gravitational acceleration. Therefore, the formalism in this paper will be different to that of Newtonian gravity or General Relativity, as a tensor treatment is not an appropriate treatment for this paper.

A schema is an outline of a model of a complex reality to assist in explaining this reality. The work of various researchers [6] in the gravity field can be presented by a conceptual formalism referred to as source-field-effect schema. The source-field-effect schema corresponds to the mass-gravity-acceleration phenomenon, respectively.

Puthoff's [22] source-field schema describes how the mass source could create a gravitational field; how General Relativity's curved spacetime could be produced by the polarizability of vacuum in the vicinity of a mass. Rueda & Haisch [22] source schema is about mass only. They discuss inertia mass, mass as a field and Higgs boson as the origin of mass. Bondi [23] suggested the possibility of a field schema not requiring mass. Bondi [23] made two observations when reviewing gravitation as a theory and suggested that mass may not be *critical* to a theory of gravitation. First, as "long as relativity is considered purely as a theory of gravitation, the inertial and passive gravitational masses do not in fact appear". This is consistent with the fact that gravitational acceleration (but not force) is independent of the mass of the object being accelerated. His second observation was that "active gravitational mass occurs for the first time as *a constant of integration* in Schwarzschild's solution" suggesting the possibility that this *constant of integration* could have other experimentally untested interpretations.

One could conjecture that mass is a proxy for number of quarks and therefore a proxy for quark interaction as the source of gravitational fields. Bondi did not explicitly say it, but maybe one should look into other mechanisms for gravitational field sources.

Hooft [24] takes another step in Bondi's direction with his source-field schema. He states that the "absence of matter no longer guarantees local flatness" that the absence of mass does not guarantee that acceleration will not be present. In effect the field is being disengaged from its source. Wagoner [25] describes a local-field schema, how a gravitational field "emerges from a local analysis" leading to a broad class of metric theories. Solomon's [6] schema proposed a different local analysis, one where local field distortions lead to a local field acceleration function, $g = w^2$, thereby providing a mathematical solution to Hooft's [24] assertion that "absence of matter no longer guarantees local flatness".

Having [2] proposed contemporary physics can be categorized by three types of particles, inelastic & point-like (quantum theory), tensile (strings) and compressive, contemporary physics recognizes three types of schema, (i) General Relativity or a geometric surface schema, (ii) Quantum Mechanics or inelastic point like elementary particle based schema, and (iii) String Theories or tensile strings based schema. This paper proposes a fourth type of schema that is closer to General Relativity but with compressive elementary particles.

Using General Relativity's [21] separation vectors this schema approach is presented by equations (2) to (6). Equation (2) presents the standard z-direction separation vector as a function of gravitational mass m, and gravitational constant G at a distance r from the source. Gravitational acceleration g can be defined in terms of separation vectors by equation (3). This three-part schema can be described as, i) the mass source or equation (6), ii) the field or equation (5), and iii) the field effect or acceleration, equation (3).

$$\frac{d^2\xi^z}{dt^2} = 2\frac{Gm}{c^2 r^3}\xi^z \tag{2}$$

$$g = f\left(\frac{d^2\xi^z}{dt^2}\right) \tag{3}$$

$$\frac{d^2\xi^z}{dt^2} = \Omega\xi^z \tag{4}$$

$$\frac{d^2\xi^z}{dt^2} = h\left(\xi^z\right) \tag{5}$$

$$\Omega = 2\frac{Gm}{c^2 r^3} \tag{6}$$

It is only necessary to limit this paper to the third part, equation (3) to validate the Center of Field approach or a field-effect schema. In this paper, unlike Newtonian gravity or General Relativity [21], the importance of the shape of spacetime lies in the fact that it informs what time dilation, mass increase and length contraction are and thus their respective spatial gradients.

If General Relativity [21], models gravity as the change in the shape of spacetime, the curving of spacetime to cause this effect of gravity, one could propose an equivalent shape change on a *non-point sized particle*; that the change in the shape of spacetime in the local region of the particle is mirrored by an identical change in the shape of the particle. This is not macro body deformation due to the gravitational gradient[ii] but particle-level deformation due to space contraction, time dilation and mass increase. The resulting deformation of the particle's shape is evidenced as a shift in the center of fields of its mass-volume 'field'.

This is a logical extension of the inertia Lorentz-Fitzgerald transformations $\Gamma(v)$, equation (7), and the Newtonian non-inertia gravitational field transformation $\Gamma(a)$, equation (8),

$$\Gamma(v) = 1 / \sqrt{(1 - v^2 / c^2)} = x_0 / x_v = t_v / t_0 = m_v / m_0 \qquad (7)$$

$$\Gamma(a) = 1 / \sqrt{(1 - 2GM / rc^2)} = x_0 / x_a = t_a / t_0 = m_a / m_0 \qquad (8)$$

Or in the generic form, the environmental transformation $\Gamma(e)$,

$$\Gamma(e) = x_0 / x_e = t_e / t_0 = m_e / m_0 \qquad (9)$$

Solomon [6 & 9] proposed that this mass-volume field deformation was due to the *internal* effects of the Newtonian non-inertia transformations $\Gamma(a)$, present in the local region of the *external* gravitational field such that the spacetime transformations $\Gamma_{s(x,y,z,t)}$ are concurrently reflected as particle transformations $\Gamma_{p(x,y,z,t)}$ or,

$$\Gamma_{p(x,y,z,t)} = \Gamma_{s(x,y,z,t)} \qquad (10)$$

The utility of equation (10) is that it explains why the gravitational field passes through all matter. Further, using the Newtonian non-inertia transformations $\Gamma(a)$ one can now replace the right hand side separation vector function in equation (3) with mass and volume spatial gradients per equation (10).

$$g = f(\Gamma(a)mass, \Gamma(a)volume) \qquad (11)$$

2.3 Gravitational Field Effect

In this section, the parameters used to determine the Center of Field C_F or specifically the Center of Mass C_M, are the spatial gradients of mass and volume. In the absence of any transformations, the center of mass C_M of the mass-volume field of a spherical[iii] particle is given by equation (12),

$$C_M = \int_{-L}^{+R} xy^2\rho_v dx \left/ \int_{-L}^{+R} y^2\rho_v d\text{:} \right. \tag{12}$$

Under the influence of a gravitational gradient, equation (13), the particle would deform such that its near-side (to the gravitational source) would be flatter than its far-side, resulting in an asymmetrical ovoid-like deformation or $|L| \neq |R|$. Simultaneously, the mass of this particle should obey an equivalent transformation, being denser on the near- than on the far-side per equation (13). In a gravitational field Φ the moments $M_{\Phi i}$ of the mass $m_{\Phi i}$ of slice i with a non-linear mass density behavior $\varrho_{\Phi i}$ is given by equation (14) and the particle's center of mass $C_{M\Phi}$ in the gravitational field Φ is given by equation (15).

$$\Gamma(x) = 1/\sqrt{1 - 2GM/(r + x)c^2} \tag{13}$$

$$M_\Phi = \int_{-L}^{+R} \left[\int_0^x \frac{dx}{\Gamma(x)} \right] \pi y_i{}^2 \rho_{\Phi i} dx \tag{14}$$

$$C_{M\Phi} = \int_{-L}^{+R} \left[\int_0^x \frac{dx}{\Gamma(x)} \right] y^2 \Gamma(x) dx \left/ \int_{-L}^{+R} y^2 \Gamma(x) dx \right. \tag{15}$$

Assume that mass density ϱ_v across the particle is uniform and constant. When inertia motion is present $(0 \leq |v| \leq c$ and $|a| = 0)$ the transformation based deformation is symmetrical, and equation (15) degenerates to equation (12) with $|L| = |R|$, and evaluates to zero. That is, when acceleration is not present there is no shift in the Center of Field; equation (15) is the standard center of mass equation (12) modified to handle the non-linearity introduced by gravitational spacetime deformation.

It should be note that equation (15) does not have an analytical solution, and a numerical model [6 & 9] was built to solve this equation. Table 1 shows 10 of 1,190 numerical modeling results that led to the discovery of the massless formulae for gravitational acceleration, equation (16),

$$g = c^2 dt / dr = \tau.c^2 \tag{16}$$

Where dt is the change in time dilation and dr the change in distance across a particle of size dr. Solomon [6 & 9] showed that both the linear Lorentz FitzGerald transformation $\Gamma(v)$ and the Newtonian non-linear

19

Planet	Gravitational Acceleration, g (m/s²) $g = GM/r^2$	Shift in the Center of Mass (m) Numerical Integration Results Equation(15)	Time Dilation (s) $t_{x=0}$ Equation (8)	Change in Time Dilation Across the Particle $\delta t = t_{i=-L} - t_{i=+R}$	Particle Size in the Gravitational Field $S_Z = \sum_{i=-L}^{i=+R} x_i$
Pluto	0.6054524	1.6904307E-41	1.00000000000007740	6.7298299E-29	9.9950000E-12
Mercury	4.0235485	1.1233798E-40	1.00000000000109230	4.4723246E-28	9.9950000E-12
Mars	3.8205065	1.066902E-40	1.00000000000144100	4.2466358E-28	9.9950000E-12
Uranus	8.7588541	2.4454830E-40	1.00000000002504600	9.7357940E-28	9.9950000E-12
Venus	8.8738716	2.4775960E-40	1.00000000000599320	9.8636402E-28	9.9950000E-12
Saturn	10.5630450	2.9492154E-40	1.00000000007040030	1.1741220E-27	9.9950000E-12
Earth	9.8028931	2.7369800E-40	1.00000000000695870	1.0896282E-27	9.9949999E-12
Neptune	11.2651702	3.1452496E-40	1.00000000003095940	1.2521658E-27	9.9950000E-12
Jupiter	24.8686161	6.9433487E-40	1.00000000019756420	2.7642398E-27	9.9949998E-12
Sun	280.3020374	7.8260679E-39	1.00000002151965680	3.1156754E-26	9.9949785E-12

transformation $\Gamma(a)$ have a deeper consistency. The escape velocity time dilation is the same as the Lorentz FitzGerald velocity time dilation. Therefore time dilation is a representation of velocity. dt/dr, the spatial gradient of time dilation, represents the spatial gradient of velocity in space or the Non Inertia (Ni) Field. See Fig. 1. As shown in Table 2, equation (16) is correct for all natural gravitational fields.

The *Ni field* is defined as a field that exhibits acceleration along a spatial gradient of latent or real velocities. Fig. 1 illustrates the *Ni field* of four velocity vectors v_1, v_2, v_3 and v_4 and their associated time dilations.

$$v_1 > v_2 > v_3 > v_4 > \tag{17}$$

$$g = f(dv/dr) = c^2\,dt/dr = \tau c^2 \tag{18}$$

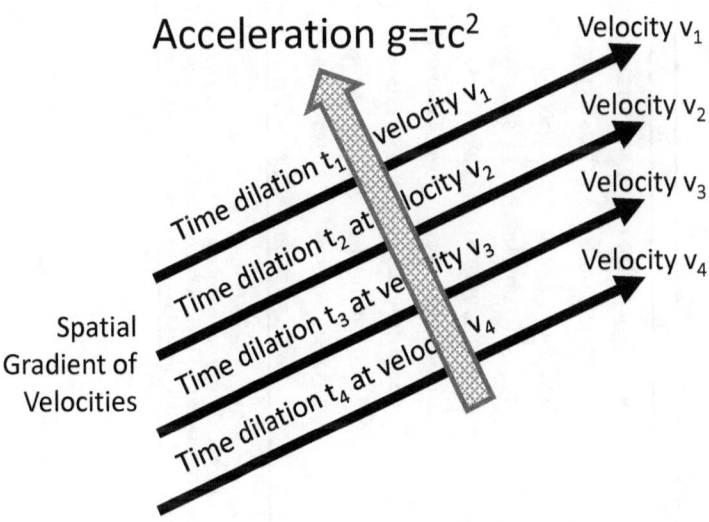

Fig. 1, A basic non-inertia field

TABLE (2). Gravitational acceleration values using Newtonian equation and equation (15), for a particle diameter of 10^{-11} m.

Heavenly Body	Gravitational $g=GM/R^2$ (m/s^2)	Acceleration $g=\tau c^2$ (m/s^2)	Error in g PPM (Parts Per Million)	Change in Time Dilation Transformation Across Particle	Gravitationally Distorted Distance (m)
Earth	9.8028931022465580	9.8028931229304500	-0.002087636085405	1.0901735980338350E-27	9.994999993044690E-12
Jupiter	24.868616073880500	24.868617547825600	-0.059269286588270	2.7656233127821500E-27	9.994999802534500E-12
Mars	3.8205064522258500	3.8205064539101700	-0.000432315179401	4.2487612761240400E-28	9.994999985596600E-12
Mercury	4.0235484580757800	4.0235484593943100	-0.000327701863842	4.4745630176813100E-28	9.994999990820000E-12
Neptune	11.265170220787000	11.265170325416100	-0.009287836549597	1.2527925259045000E-27	9.994999960560200E-12
Pluto	0.6054524007561430	0.6054524007702150	-0.000023241154340	6.7331981932573400E-29	9.994999992560E-12
Saturn	10.563045030532200	10.563045253624700	-0.021120092215353	1.1747096324395700E-27	9.994999296348900E-12
Sun	280.30203737103000	280.30384692092000	-6.455910938663970	3.1172348020403100E-26	9.994978491149290E-12
Uranus	8.7588540780329700	8.7588541434853400	-0.007513810986813	9.7406667599414200E-28	9.994999974966480E-12
Venus	8.8738715534578700	8.8738715694128000	-0.001797966844070	9.8685769363557400E-28	9.994999994009770E-12

Notes: (a) Though the data presented is only to 15 decimal places, all numerical analyses were conducted to 250 significant digits using G of 6.67259x10^{-11}. (b) The numerical results validates equation (16), that gravitational acceleration can be derived without reference to its mass source as the errors between Newtonian g (column 2) and Ni Field g (column 3) is <|6.5| ppm, (c) Combining all recent experimental results [29] provides measured range of between 6.671x10-11 to 6.676x10^{-11} or a mean of 6.6738x10^{-11} and standard deviation of 0.0012x10^{-11} that the true G is in this range. (d) These near field gravity probe G measurements are a good example of precision versus accuracy and proof of the presence of systematic errors

2.4 Electromagnetic & Mechanical Accelerations

This section, proposes how the spatial gradient of velocities or *Ni fields* govern acceleration in mechanical and electromagnetic fields. A mechanical field would be a field of velocities produced by the motion of some mechanical structure.

In current electromagnetic theory, the acceleration *a* is a function of tangential velocity *v*, electron charge *q*, magnetic field *B* and electron mass *m* defined by equation (19). The centripetal method is the standard mechanics method, acceleration *a* is a function of the tangential velocity *v* and the radius of circular motion *r* defined by equation (20). With angular velocity *ω*, tangential velocity *v* is given by equation (21).

$$a = q(v \times B)/m \tag{19}$$

$$a = v^2/r \tag{20}$$

$$v = \omega r \tag{21}$$

In the mechanical *Ni field* method, column 7 of Table (3), the tangential velocity is converted to time dilation, and acceleration is calculated using equations (7) and (18) where *dt* is time dilation at velocity *v minus 1* (the time dilation at center of circular motion with zero velocity), and *dr* is the radius of the circular motion. The new electromagnetic *Ni field* model results are given in column 9 of Table (3). The force field equation (18) agrees with current electromagnetic theory equation (19), centripetal mechanics equation (20) and therefore correct for mechanical, electromagnetic and gravitational forces i.e. equation (18) is the universal description of acceleration for non-nuclear forces. The new *Ni field* method agrees with other methods except in the last row where particle size is $> 10^{-3}$m. This is because the distance between the two velocities are great enough to be the *effective* or *averaged* acceleration over a large distance.

The magnetic field, Fig. 2, causes the upward moving charged particle with velocity *v* to rotate about its center of rotation with an angular rotation *ω* and a path radius *r*. By convention the particle's negatively charged electric field *E* of some radius *dr* is pointing into the particle. Due to the arched path in the magnetic field, the particle will have a small change in velocity *dv* such that the field's *left-* and *right*-side velocities along the path radius are given by equations (22) and (23), respectively.

$$v\text{-}dv = \omega(r\text{-}dr) \tag{22}$$

$$v\text{-}dv = \omega(r\text{+}dr) \tag{23}$$

In other words, the $\pm$ *dv* shows that just as the electron's path rotates about an external center of rotation of radius *r*, the electron's electric field is itself, rotating in an anti-clockwise manner about internal center of rota-

TABLE (3). Acceleration calculated using current electromagnetic theory equation (19), centripetal method equation (20), mechanical *Ni field* method equation (18), and the electromagnetic *Ni field* model equation (31).

Angular Velocity	Path Radius (m)	Tangential Velocity (m/s)	Particle Size (m)	Acceleration					
				Centripetal Method	Electro-Magnetic Theory	Mechanical Ni Field Method	Mechanical Ni Field Error	Electromagnetic Ni Field Method	Electro-magnetic Ni Field Error
				Equation (20)	Equation (31)	Equations (18)	(18)-(20)	Using Eqn. (31)&(18)	(18)-(29)
68	9.85	669.8	1.00E-30	45,546.40	45,546.40	45,546.40	2.4033E-13	45,546.4000003409	-3.4E-07
127	4.07	516.89	1.00E-27	65,645.03	65,645.03	65,645.03	4.1720E-14	65,645.0300002925	-2.9E-07
49	5.01	245.49	1.00E-24	12,029.01	12,029.01	12,029.01	3.2700E-15	12,029.0100000120	-1.2E-08
98	6.13	600.74	1.00E-21	58,872.52	58,872.52	58,872.52	6.0000E-16	58,872.5200003546	-3.5E-07
148	4.75	703	1.00E-18	104,044.00	104,044.00	104,044.00	1.7090E-12	104,044.000000858	-8.6E-07
116	0.42	48.72	1.00E-15	5,651.52	5,651.52	5,651.52	9.6100E-15	5,651.5200000002	-2.2E-10
96	0.79	75.84	1.00E-12	7,280.64	7,280.64	7,280.64	3.3797E-12	7,280.6400000007	-6.8E-10
2	1.17	2.34	1.00E-09	4.68	4.68	4.68	6.0083E-12	4.680000000	-5.1E-14
74	1.86	137.64	1.00E-06	10,185.36	10,185.36	10,185.36	2.1640E-14	10,185.3600000812	-8.1E-08
170	4.64	788.8	1.00E-03	134,096.00	134,096.00	134,096.00	7.4337E-13	134,096.000413369	-4.1E-04

tion, with radius dr. This anti-clockwise rotation of the electric field is due to the interaction of the electron's radial electric field with the external magnetic field per the Right Hand Rule. Solving equations (19), (20) and (21), gives (24) and (25). By equations (22) and (23) the electron rotation about its own center must be equal to the rotation about its arched path. Therefore, the electron field velocity dv is given by equation (26).

$$v = (q/m)Br \tag{24}$$

$$\omega = (q/m)B \tag{25}$$

$$dv = (q/m)Bdr \tag{26}$$

Converting $v \pm dv$ at any point in the electric field into time dilations using equation (7) gives equation (27) as t_0 at the center of rotation is 1. Therefore, the difference in time dilations across any two points in the electric field, say points 1 & 2, is given by (28) or (29).

$$t_v = t_0 / \sqrt{1 - (v+dv)^2/c^2} = 1/\sqrt{1 - (v+dv)^2/c^2} \tag{27}$$

$$dt = 1/\sqrt{1-(v+dv_1)^2/c^2} - 1/\sqrt{1-(v+dv_2)^2/c^2} \tag{28}$$

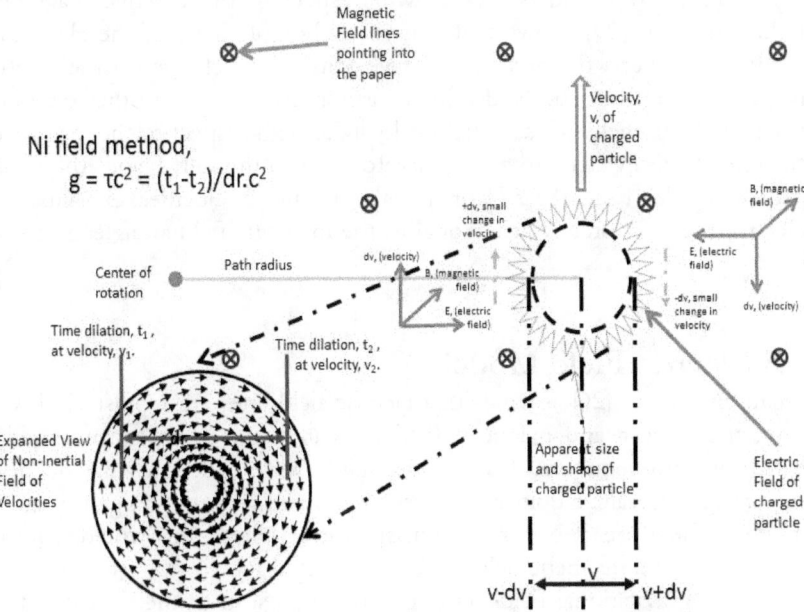

Fig. 2, Charged particle moving in a magnetic field forms a non-inertia field

$$dt = 1/\sqrt{1-(v+(q/m)Bdr_1)^2/c^2} - 1/\sqrt{1-(v+(q/m)Bdr_2)^2/c^2} \qquad (29)$$

For the numerical modeling, an infinitesimally small piece of the spherical field, approximates a flat surface of $(q/A)/(2\varepsilon_0)$. Thus the small change in velocity dv is given by equation (30), and its vertical component orthogonal to the horizontal electric field component E_x ($E\cos\theta$) is equation (31).

$$dv = (8\pi\varepsilon_0/m)BEdr^3 \qquad (30)$$

$$dv = (8\pi\varepsilon_0/m)(E\cos\theta)Bdr^3 \qquad (31)$$

The charged particle's electric field acceleration is presented in column 9 of Table (3). However, a careful examination of the results shows that the sign is incorrect. It turns out that the Left Hand Rule provides the correct magnitude *and sign* of the electron acceleration. The inference is that, the Right Hand Rule is the correct rule for electromagnetic field propagation when the magnetic field is *intrinsic* to the motion; but when the magnetic field is *external or foreign* to the electric field the Left Hand Rule is correct. Thus, with respect to electron rotation the small change in velocity dv is reversed such that the field's *right-* and *left*-side velocities along the path radius are given by equations (22) and (23), respectively. Or the moving electron's electric field is locked with respect to the external magnetic field. Equation (29) shows that it is the spherical shape of the electrical field that converts the perpendicular velocity of the charged particle with respect to the magnetic field into an orthogonal force. All other orientations of the particle's electric field velocity dv with respect to the magnetic field negate or evaluate to zero. Therefore, given the Left Hand Rule, this process model, using *Ni fields* or spatial gradient of velocities, explains *why* electromagnetic force is orthogonal to the magnetic field and electron velocity.

2.5 Electric Field Model

Quantum theory [26] assumes that electric field lines are repulsive. However, the electric and magnetic field lines in transverse electromagnetic waves are not repulsive. These form stable wave functions that do not require an equivalent Poincaré stresses treatment as with charged particle models. Therefore, if Nature is consistent everywhere, this would require that electric and magnetic field lines are not repulsive.

As with gravitational fields, in electric fields, a shift of the C_E either towards or away is representative of attraction or repulsion, respectively. In this section, the shift in the Center of Field C_F is handled differently for the two scenarios considered.

Two charged particles: The field property of interest is the *spatial gradient* $\delta E_i / \delta d_i$ at location i of the electric field E. This parameter is used to determine the shift[iv] in the Center of Field C_E of a charged particle's electric field as non-linear deformation is introduced into its field by the other charged particle's electric field.

Parallel plates: Charged parallel plates introduce linear deformations to a charged particle's electric field as the *spatial gradient* of the parallel plates' electric field is zero. Therefore, to evaluate the charged particle's electric field's Center of Field C_E shift, the electric field E itself, is the field property P of interest.

Per the transverse wave, assuming that electric field lines are not repulsive, a straight forward summative treatment should be sufficient,

Two charged particles: The summing of the *spatial gradient* of the electric field lines $\delta E_{i,C} / \delta d_i$ or change in the combined electrical field strength, $\delta E_{i,C}$, divided by the change in that distance, δd_i, at a location i, is the basis for calculating the electrical field's shift in its Center of Field C_E,

$$C_E = \sum \left(\delta E_{i,c} / \delta d_i \right) x_i / \sum \left(\delta E_{i,c} / \delta d_i \right) \tag{32}$$

Parallel plates: The summing of the parallel plates electric field lines $E_{i,p}$ at a location i, is the basis for calculating the charged particle's electrical field's Center of Field C_E,

$$C_E = \sum E_{i,P} x_i / \sum E_{i,P} \tag{33}$$

The convention for summing either the electric field lines E_i or the spatial gradient of these electric fields, $\delta E_i / \delta d_i$ is the same as that of conventional electric field lines:

1. Field lines point in the direction of the greater negative.
2. Use the Cartesian convention that right pointing is positive and left pointing is negative, and similarly with up & down and ahead & behind, respectively.
3. Unlike directions are subtracted and like directions are added.
4. If necessary resolve electric field lines into their x-, y- & z-directions.

Effectively, this can be restated in experimental design terms as,

Null Hypothesis: Electric field lines are not repulsive in a non-linear manner, and therefore summative in a linear manner. For *both* scenarios, the two charged particles & the parallel plates, there are exact mathematical solutions that govern velocity and acceleration with respect to the shift in C_E.

Alternate Hypothesis: Electric field lines are repulsive in a non-linear manner such that, they cannot be summative in a linear manner. Therefore, for *both* cases, mathematical solutions for repulsive forces cannot be determined by linear summation.

For parallel plates the electric field $E_{i,P}$ at a distance x_i is determined by the voltage V between the plates separated by a distance d and is a constant for all i. Thus, it does not introduce non-linear deformations to the charged particle's electric field,

$$E_{i,P} = V/d \qquad (34)$$

Therefore, the combined electric fields $E_{i,C}$ for a charged particle's Q electric field $E_{i,Q}$ between parallel plates is the straight sum of the electric fields of the parallel plates $E_{i,P}$ and of the charged particle $E_{i,Q}$,

$$E_{i,C} = E_{i,P} + E_{i,Q} \qquad (35)$$

For any charged particle Q the gradient of the electric field is the difference of the electric field strengths $\delta E_{i,Q}$ at distances x_i and x_{i+1} from the center of this charged particle, given permittivity ε_0,

$$E_{i,Q} = (1/4\pi\varepsilon_0)Q/x_i^2 \qquad (36)$$
$$E_{i+1,Q} - (1/4\pi\varepsilon_0)Q/x_{i+1}^2 \qquad (37)$$
$$\delta E_{i,Q} = E_{i,Q} - E_{i+1,Q} \qquad (38)$$

Similarly, for two charged particles Q_L on the left and Q_R on the right, the spatial gradients of the combined fields, $\delta E_{i,C}/\delta x_i$, is

$$\delta E_{i,C}/\delta x_i = \delta E_{i,QL}/\delta x_i + \delta E_{i,QR}/\delta x_i \qquad (39)$$

Or in the limit

$$dE_{i,C}/dx_i = dE_{i,QL}/dx_i + dE_{i,QR}/dx_i \qquad (40)$$

A numerical model was constructed to implement equations (35) and (40). A best fit regression model was applied to determine the C_E functions. The modeling gave the following equations for velocity v and acceleration a.

Parallel plates: The velocity v of a charged particle between parallel plates is given by,

$$v = k(Q/d)\sqrt{C_E} \qquad (41)$$
$$k = -3.7676 \times 10^{20}\sqrt{(m_e/m_p)} \qquad (42)$$

where k is a function of electron mass m_e and particle mass m_p.

Two charged particles: The right particle Q_R introduces non-linear deformations to the left Q_L particle's electric field. Therefore, for a particle separation of 1.4090x10⁻¹⁴m, the acceleration of the left particle Q_L is given by,

$$a = C_E Q_R [k_1(Q_R/Q_E) + k_2] + [(Q_R/Q_E)^2 k_3 + k_4] \tag{43}$$

and C_E is given by

$$C_E = k_5(Q_R/Q_E) + k_6 \tag{44}$$

Table 4 provides the value of the constant terms with,

k_5 = $9.2015x10^{-14}$
k_6 = $9.2547x10^{-15}$

Per equation (43), Fig. 3 shows the ratio of acceleration to C_E, is smooth continuous surface.

Note, that k_1 and k_2 are the same for both attraction and repulsion. The differences in k_3 and k_4 have no significant effect on acceleration.

Table 5 provides the errors observed between classical calculations and the spatial gradient model, equation (43) for like charges.

The objective to demonstrate the Null Hypothesis that both summative equations do exists is accomplished. Therefore, with R^2 at or greater than 99.9999%, there was little point in searching for a better fit model to reduce the first row errors, using weighted sum of squares regression optimization.

The Null Hypothesis is accepted, that for electric fields, the Center of Fields C_E, do provide an alternative mechanism for determining velocities and accelerations. Therefore, electric field lines are not repulsive in a non-linear manner.

This finding is consistent with the behavior of electric field lines in transverse electromagnetic waves.

Similarly, by analogy, one can conclude that the same is true for magnetic monopole lines, with the just the monopole and constant terms taking on different values.

Therefore, by extension of the magnetic monopoles, magnetic field lines are also not repulsive in a non-linear manner. This conclusion is vindicated by the non-repulsive behavior of magnetic field lines in the transverse

Table 4: Value of constant terms for equation (43)		
Constant Terms	Repulsion	Attraction
k_1	$5.0544x10^{13}$	$5.0544x10^{13}$
k_2	$2.9732x10^{10}$	$2.9732x10^{10}$
k_3	$4.8280x10^{-1}$	$4.7874x10^{-1}$
k_4	$3.9298x10^{-3}$	$2.8301x10^{-3}$

Table 5: Error (%) between Classical and Spatial Gradient Models

Q_L/Q_E \ Q_R/Q_E	0.33333	0.66667	1	1.33333	1.66667	2	2.33333	2.66667	3	3.33333	3.66667	4	4.33333	4.66667
0.33333	-0.578%	-0.260%	-0.147%	-0.094%	-0.071%	-0.070%	-0.085%	-0.114%	-0.157%	-0.213%	-0.281%	-0.361%	-0.453%	-0.557%
0.66667	-0.225%	-0.105%	-0.057%	-0.028%	-0.008%	0.005%	0.013%	0.018%	0.019%	0.016%	0.010%	0.002%	-0.010%	-0.025%
1	-0.107%	-0.056%	-0.032%	-0.016%	-0.003%	0.007%	0.016%	0.022%	0.027%	0.031%	0.033%	0.034%	0.033%	0.031%
1.33333	-0.049%	-0.032%	-0.021%	-0.011%	-0.003%	0.005%	0.011%	0.017%	0.022%	0.027%	0.030%	0.033%	0.035%	0.037%
1.66667	-0.014%	-0.018%	-0.014%	-0.009%	-0.004%	0.001%	0.007%	0.012%	0.016%	0.020%	0.024%	0.027%	0.030%	0.033%
2	0.010%	-0.008%	-0.010%	-0.008%	-0.005%	-0.001%	0.003%	0.007%	0.011%	0.015%	0.018%	0.021%	0.024%	0.027%
2.33333	0.026%	-0.002%	-0.007%	-0.008%	-0.006%	-0.003%	0.000%	0.003%	0.006%	0.010%	0.013%	0.016%	0.019%	0.022%
2.66667	0.039%	0.003%	-0.005%	-0.007%	-0.007%	-0.005%	-0.003%	0.000%	0.003%	0.006%	0.009%	0.011%	0.014%	0.017%
3	0.049%	0.007%	-0.004%	-0.007%	-0.007%	-0.006%	-0.005%	-0.003%	0.000%	0.002%	0.005%	0.007%	0.010%	0.012%
3.33333	0.056%	0.010%	-0.002%	-0.007%	-0.008%	-0.008%	-0.006%	-0.005%	-0.003%	-0.001%	0.002%	0.004%	0.006%	0.009%
3.66667	0.063%	0.013%	-0.001%	-0.007%	-0.009%	-0.009%	-0.008%	-0.007%	-0.005%	-0.003%	-0.001%	0.001%	0.003%	0.005%
4	0.068%	0.015%	0.000%	-0.007%	-0.009%	-0.010%	-0.009%	-0.008%	-0.007%	-0.005%	-0.003%	-0.001%	0.000%	0.002%
4.33333	0.073%	0.017%	0.000%	-0.006%	-0.009%	-0.010%	-0.010%	-0.009%	-0.008%	-0.007%	-0.005%	-0.004%	-0.002%	0.000%
4.66667	0.076%	0.018%	0.001%	-0.006%	-0.010%	-0.011%	-0.011%	-0.010%	-0.009%	-0.007%	-0.006%	-0.006%	-0.004%	-0.002%

electromagnetic wave. As a result, non-linear repulsive electric and magnetic field lines are falsifiable, and not allowed in Nature.

The four unintended consequences of this Null Hypothesis are:

Poincaré stresses do not exists: As electric field line are not repulsive on the 'surface' of the charged particle, Poincaré stresses [26] are not required to counterbalance these repulsive forces. Thus, proving to be a new approach to developing new theories, per Steinhardt & Esftathiou's [11].

Experimental Error: The concept of electric and magnetic field lines are due to experimental error of our methods. In magnetic field lines, the use of discrete iron filings causes the effect of lines. In a magnetic field, the first filing causes the greatest field deformation towards itself. As the strongest magnet in the field, it attracts the next filing aligned along its *own* north-south alignment. This process is repeated until one observes a line of filings. Thus, the perception of lines. However, with ferromagnetic liquids, one observes conic-like structures, not lines.

A New Field Interpretation: Therefore, it is suggested, that in the absence of field lines, electric and magnetic fields are volume structures whose structure needs to be consistent with the electromagnetic transverse wave's infinitely thin electric and magnetic fields.

Unified Theory of Everything (UTE): The process models approach provides an alternative avenue to developing UTEs by investigating the elementary *process* behaviors.

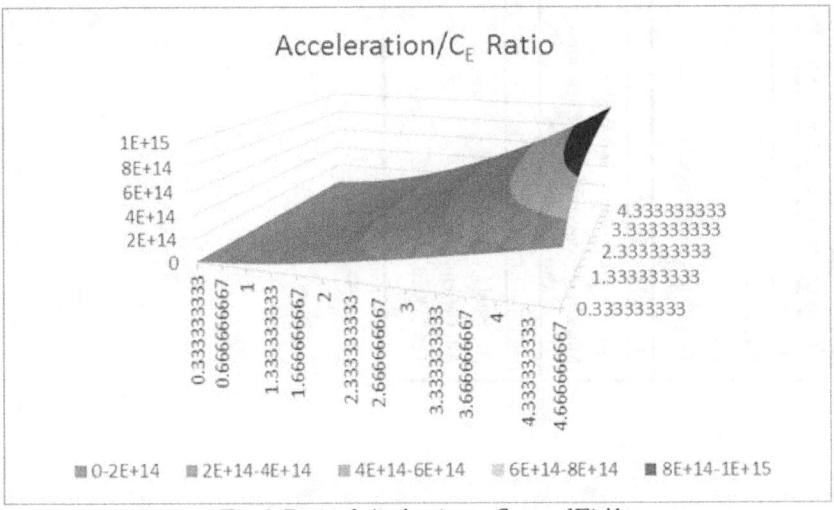

Fig. 3, Ratio of Acceleration to Center of Field

Table 6: Proton & Deuteron Process Models

Standard Data	Proton	Deuteron	Avg. or Sum
Particle Mass (kg)	1.672622×10^{-27}	3.343583×10^{-27}	
RMS Charge Radius	8.775000×10^{-16}	2.142400×10^{-15}	
Total Quark Rest Masses In Particle	9.4	21.3 MeV	
	1.6757×10^{-29}	3.79707×10^{-29} kg	
Process Model	**Proton**	**Deuteron**	**Avg. or Sum**
Quark/Particle Mass Ratio (M_Q/M_V)	1.001842×10^{-02}	1.135629×10^{-02}	1.070827×10^{-02}
Mass Ratio converted to LFT velocity	$299{,}777{,}413$	$299{,}773{,}126$	$299{,}775{,}269$
Error % (wrt average)	-0.000715%	0.000715%	$0.00143 0\%$
Adjusted Effect	**Proton**	**Deuteron**	
Adjusted LFT velocity^2	$8.986521 \times 10^{+16}$	$8.986521 \times 10^{+16}$	
Adjusted mass ratios	1.070827×10^{-02}	1.070827×10^{-02}	
Total Quark Masses	1.791088×10^{-29}	3.580398×10^{-29} kg	
	10.047	20.085 MeV	

2.6 Up & Down Quark Masses

The elimination of Poincaré stresses substantially simplifies the particle model, and per Steinhardt & Esftathiou's [11] an new approach is warranted. Since the proton's quarks masses are about 1% of the proton mass, one could use the difference between proton mass (939.56 MeV[i]) and its total standard quark masses (up quark of 2.3{+0.7/-0.5} MeV and down quark of 4.8 {+0.5/-0.3} MeV[ii]) to investigate a simplified particle process model.

Experimenters [27] have not been able to measure quark masses directly. Theorists are unable to calculate them with great precision, until Davies et al [18] who have suggested 2.01 ±0.14 & 4.79±0.16 MeV, respectively. However, by examining proton and neutron masses, Jefferson Labs[iii], have suggested that both quarks ought to have similar masses.

As Poincaré stresses [26] don't exists and electric field lines are not repulsive, the simplest process model, the Quark Velocity model, would require quarks 'orbiting' inside a proton at the RMS charge radius of the proton of 8.775×10^{-16} m[viii] and for comparison a deuteron of mass 1,875.61 MeV with an RMS charge radius of 2.1424×10^{-15} m[7].

Unlike the Standard Model's [27] infinity of gluons, quarks & antiquarks, from the perspective of mass, the Quark Velocity model, assumes that only 2 up & 1 down quarks are 'orbiting' in the proton and 1 up & 2 down quarks are 'orbiting' in the deuteron. Then 'orbiting' or tangential velocity could be determined from the ratio of masses using Lorentz FitzGerald Transformation (LFT) or equation (7).

Table 6 shows that the difference between proton & deuteron quarks' 'orbital' velocities is only 0.001430%. This apparent consistency is a very surprising outcome. Using the mean velocity for a common mass ratio of 1.070827×10^{-02} with standard & Davies et al [28] quark masses as a starting point to determine the process model quark masses, Table 7 summarizes the results.

The process model results are quite stable, only 3.57% difference. However, the neutron mass error is 3.80% when using values derived from Davies et al [28] as starting masses for these iterations. Therefore, iterating to minimize all mass errors (proton 938.272046 MeV, neutron 939.565379 MeV & deuteron 1,875.612860 MeV) gives up and down quark masses of 3.343167638 and 3.356973524 MeV, respectively. This agrees with the Jefferson Lab's estimate that quark masses have to be similar.

Of course, much work is required to develop these process models, but without Poincaré stresses, these simpler models do show promise.

Table 7: Neutron mass derived from starting		
Starting Quark Masses	Standard Model	Davies *et al*
Up Quark	2.3	2.01
Down Quark	4.8	4.79
Iterative Solution		
Up Quark	3.352	3.233
Down Quark	3.342	3.223
Neutron Mass	937.340813	903.880780
Error	0.24%	3.80%

2.7 Conclusion

This paper has shown that the Spatial Gradient Center of Fields is the common thread running through, gravitation, mechanical, electric, magnetic, electromagnetic, velocities and accelerations. Thereby, providing an alternative to Relativistic, String & Quantum (RSQ) theories. As a consequence both Poincaré stresses and non-linear repulsive electric & magnetic field lines are falsifiable and will simplify future particle models. With the removal of Poincaré stresses, this paper has proposed a simple manner of estimating up & down quark masses.

References:

[1] Solomon, B. T.: Empirical Evidence Suggest A Need For A Different Gravitational Theory, in the proceedings of the *100 Year Starship Study Public Symposium (100YSS,)*, 2013.

[2] Solomon B.T.: New Evidence, Conditions, Instruments & Experiments for Gravitational Theories, Journal of Modern Physics, Special Issue on Gravitation, Astrophysics and Cosmology, Vol. 8A, 2013, August 2013.

[3] Solomon, B. T.: Empirical Evidence Suggest A Need For A Different Gravitational Theory, *American Physical Society (APS) April Conference, Denver*, 2013

[4] Solomon B.T.: An Introduction to Gravity Modification: A guide to using Laithwaite's and Podkletnov's experiments and the physics of forces for empirical results. Universal Publishers, Boca Raton, 2nd Edition, May 2012.

[5] Solomon, B. T.: Non-Gaussian Radiation Shielding, *100 Year Starship Study Public Symposium (100YSS,)*, 2011.

[6] Solomon, B.T.: Gravitational Acceleration Without Mass And Noninertia Fields, Physics Essays, Vol. 24, 327, 2011. [Phys. Essays **24**, 327 (2011)]

[7] Solomon, B. T.: Reverse Engineering Podkletnov's Experiments, in the proceedings of the *Space, Propulsion & Energy Sciences International Forum (SPESIF-11)*, Edited by Glen A Robertson, Physics Procedia, Elsevier Science.

[8] Solomon, B. T.: Non-Gaussian Photon Probability Distributions, in the proceedings of the *Space, Propulsion & Energy Sciences International Forum (SPESIF-10)*, Edited by Glen A Robertson, AIP Conference Proceedings **1208**, Melville, New York, (2010).

[9] Solomon, B. T.: An Approach to Gravity Modification as a Propulsion Technology, in the proceedings of the *Space, Propulsion & Energy Sciences International Forum (SPESIF-09)*, Edited by Glen A Robertson, AIP Conference Proceedings **1103**, Melville, New York, (2009).

[10] Solomon, B. T.: An Epiphany On Gravity, Journal of Theoretics, Vol. 3-6, 2001.

[11] Efstathiou, G., Pryke, C., Steinharrd, P., Tuttle K., Spotlight Live: Looking Back in Time - Oldest Light in Existence Offers Insight into the Universe, (video blog) The Kavli Foundation, 02/18/2015. http://www.kavlifoundation.org/science-spotlights/looking-back-time-oldest-light-existence-offers-insight-universe

[12] R. Nemiroff, "Bounds on Spectral Dispersion from Fer-mi-detected Gamma Ray Bursts", Phys. Rev. Lett. 108, 231103 (2012) http://dx.doi.org/10.1103/PhysRevLett.108.231103

[13] J. H. Gundlach, and S. M. Merkowitz, "Measurement of Newton's Constant Using a Torsion Balance with Angular Acceleration Feedback" Phys. Rev. Lett. 85, 2869 (2000). http://dx.doi.org/10.1103/PhysRevLett.85.2869

[14] H. V. Parks and J. E. Faller, "A Simple Pendulum Determination of the Gravitational Constant" Phys. Rev. Lett. (2010) http://xxx.lanl.gov/abs/1008.3203.

[15] Jun Luo, Qi Liu, Liang-Cheng Tu, Cheng-Gang Shao, Lin-Xia Liu, Shan-Qing Yang, Qing Li, and Ya-Ting Zhang, "Determination of the

Newtonian Gravitational Constant G with Time-of-Swing Method" Phys. Rev. Lett. 102, 240801 (2009). http://dx.doi.org/10.1103/PhysRevLett.102.240801

[16] St. Schlamminger, E. Holzschuh, W. Kündig, F. Nolting, R. E. Pixley, J. Schurr, and U. Straumann. "Measurement of Newton's gravitational constant", Phys. Rev. D 74, 082001 (2006).

[17] E. Podkletnov, "Weak gravitational shielding properties of composite bulk YBa2Cu3O7-x superconductor below 70K under e.m. field," lanl.gov, (1997).

[18] E. Podkletnov, R. Nieminen, "A Possibility of Gravitational Force Shielding by Bulk YBa2Cu3O7-x Superconductor" Physica C, 203, (1992), pp. 441-444. http://dx.doi.org/10.1016/0921-4534(92)90055-H

[19] R.C. Woods, S.G. Cooke, J. Helme, C.H. Caldwell, "Gravity Modification by High-Temperature Superconductors", in the proceedings of the 37th AIAA/ASME/SAE/ASSEE Joint Propulsion Conference & Exhibit, 8-11 July, 2001, Salt Lake City, Utah.

[20] G. Hathaway, B. Cleveland and Y. Bao., "Gravity Modification Experiments Using a Rotating Superconducting Disk and Radio Frequency Fields", Physica C, Volume 385, Issue 4, p. 488-500.

[21] Misner, C.W., Thorne, K.S., Wheeler, J.A.: Gravitation, W.H. Freeman and Company, New York, NY, 1973.

[22] *Gravitation and Cosmology: From the Hubble Radius to the Plank Scale, Proceedings of a Symposium in Honour of the 80th Birthday of Jean-Pierre Vigier*, Edited by Amoroso, R.L., Hunter, G., Kafatos, M., and Vigier, J-P., (Kluwer Academic Publishers, Boston, USA, 2002).

[23] H. Bondi, Reviews of Modern Physics, **29-3**, 423 (1957).

[24] G. Hooft, Found Phys **38,** 733 (2008).

[25] R. V. Wagoner, 26th SLAC Summer Institute on Particle Physics, **SSI 98**, 1 (1998).

[26] Feynman, R.: Feynman Lectures on Physics, Chapter 28-4.

[27] Cho, A: Mass of the Common Quark Finally Nailed Down, Science Magazine, 2 April, 2010.

[28] Davies, C. T. H., McNeile, C., Wong, K. Y., Follana, E., Horgan, R., Hornbostel, K., Lepage, G. P., Shigemitsu, J., Trottier, H.: Precise Charm to Strange Mass Ratio and Light Quark Masses from Full Lattice QCD. Physical Review Letters, 2010; 104 (13): 132003 DOI: 10.1103/PhysRevLett.104.132003

[29] Cowen, R., Quantum method closes in on gravitational constant: Cold rubidium atoms provide fresh approach to measuring Newton's big G., Nature News, 18 June 2014 (http://www.nature.com/news/quantum-method-closes-in-on-gravitational-constant-1.15427)

End Notes:

i. Does not include non-peer reviewed conference presentations between 2001 and 2009.

ii. Macro body elongation due to tidal gravity is attributed to Roger Penrose (this author could not find the reference in time for this paper). Macro bodies elongate as the body falls into a gravitational field. Let's reexamine this tidal behavior with the additional requirement that this tidal gravity property be consistent with Lorentz-FitzGerald transformations or Special Theory of Relativity. To be consistent with Lorentz-FitzGerald transformations, the atoms and elementary particles would contract in the direction of the fall. However, to be consistent with tidal gravity's elongation, the distances between atoms in the macro body has to increase at a rate consistent with the acceleration and velocities experienced by the various parts of the macro body. That is, as the atoms get flatter, the distances apart get longer. This can be named the Tidal Axiom. One suspects that this Tidal Axiom's inconsistency with the empirical evidence has led to an explosion of string theories, each trying to explain Nature with no joy.

iii. Note spherical particle shape was a convenient simplification. See Solomon [6 & 9] for more details on other shapes.

iv. In the absence of any other electric or moving magnetic fields, the Center of Fields C_E would evaluate to zero as charged particles fields have spherical symmetry. Therefore, the deformed electric field's C_E is the shift from this zero position.

v. NIST Reference on Constants, Units & Uncertainty http://physics.nist.gov/cuu/Constants/index.html

vi.   J. Beringer et al. (Particle Data Group), PR D86, 010001 (2012) and 2013 partial update for the 2014 edition, http://pdg.lbl.gov/2013/tables/rpp2013-sum-quarks.pdf

vii. Jefferson Labs have suggested that up and down quark masses should be similar http://education.jlab.org/qa/particlemass_03.html

viii. NIST Reference on Constants, Units & Uncertainty http://physics.nist.gov/cuu/Constants/index.html

ix. τ is the dimensionless ratio of the change in time dilation over that distance, per t_v/t_0 equation (7) and t_a/t_0 equation (8). Since t_0 is always 1, as a shorthand one can describe this as the time dilation at those points.

3. The Variable Isotopic Gravitational Constant

Abstract:- To determine how matter causes gravitational fields, the empirical data of 798 isotopes, consisting of RMS charge radius $r_{i,c}$, isotope masses M_i, and 703 atomic shell radius $r_{i,a}$, was analyzed to construct several very strong empirical relationships ($R^2>99.9999\%$).

It was found that the gravitational constant G, is not a constant, but a variable G_i that decreases with the isotopic mass M_i, of isotope i. Thus, the *composite gravitational constant* $G_C = k_{a,R}(\Sigma_i G_i)$ where $k_{a,R}$ is an *aggregation factor at a radial distance R*. It was shown that mass is a proxy for matter as the gravitational field of a single nuclei is not a function of its mass.

Therefore, two aggregation factors come into play when matter is accumulated in local spacetime. First, mass as an aggregation factor, determines how the quantity of matter increases the gravitational field strength. Second, $k_{a,R}$ the *aggregation factor at a radial distance R,* determines how the clustering of matter alters the gravitational field.

These aggregation factors and the isotopic gravitational constants G_i are the unknown systematic error in near field gravitational experiments. Nucleosynthesis would decrease the gravitational constant as heavier elements are formed.

The nucleosynthesis would cause the Universe's gravitational constant G_U to drop from greater than ^{1}H's $G_i = 1.777957 \times 10^{-09}$ to 4.441839×10^{-10} ($75\%\,^1$H & $25\%\,^4$He) within the first few minutes, and thus the Big Bang. Stellar gravitational constants would drop similarly depending on the amount of hydrogen remaining and heavier elements produced.

Given a nucleosynthetic decreasing composite gravitational constant G_C it was possible to review the gravitational rotational curves and infer i) two different stellar processes, ii) a galactic Main Sequence and iii) substantial changes to a galaxy's aggregation factor, k_a.

3.1 Premise

This paper reports the 12th paper in the 16-year ([1] to [11]*) investigation into the feasibility of gravity modification, proposing:

i) Non Inertia (Ni) Fields, defined as the *spatial gradient* of real or virtual velocities,

ii) Equation (1), the massless formula, for gravitational acceleration, where τ is the change in time dilation divided by the change in that distance,

$$g = \tau c^2 \tag{1}$$

iii) All macro forces (electromagnetic, electric & magnetic, mechanical, gravitational), can be model by the spatial gradient of their fields [1], and,

iv) How the isotopic gravitational constant model (this paper) causes gravitational fields, and new insights in galaxy formation.

As a consequence of iii), that electric and magnetic field lines are not repulsive, Poincaré stresses do not exist [1]. This facilitates the substantial simplification of the proton, neutron & atomic nuclei models. For example [1] with only quarks, quark masses are easily determined.

Occam's Razor, that the simpler solution is most likely the correct solution, drives this investigation for a simpler matter model that explains how matter, not mass, creates gravity.

3.2   Introduction

Solomon [6] proposed the schema approach to solving gravitational field physics. A schema is an outline of a model of a complex reality to assist in explaining this reality. The work of various researchers [6] in the gravity field can be presented by a conceptual formalism referred to as source-field-effect schema, and corresponds to the mass-gravity-acceleration phenomenon, respectively. Using General Relativity's [12] separation vectors this schema approach is presented by equations (2) to (6). Equation (2) presents the standard z-direction separation vector as a function of gravitational mass m, and gravitational constant G at a distance r from the source. Gravitational acceleration g can be defined in terms of separation vectors by equation (3). This three-part schema can be described as, i) the mass source or equation (6), ii) the field or equation (5), and iii) the field effect or acceleration, equation (3).

$$\frac{d^2\xi^z}{dt^2} = 2\frac{Gm}{c^2 r^3}\xi^z \tag{2}$$

$$g = f\left(\frac{d^2\xi^z}{dt^2}\right) \tag{3}$$

$$\frac{d^2\xi^z}{dt^2} = \Omega\xi^z \tag{4}$$

$$\frac{d^2\xi^z}{dt^2} = h\left(\xi^z\right) \tag{5}$$

$$\Omega = 2\frac{Gm}{c^2 r^3} \tag{6}$$

If General Relativity [12], models gravity as the change in the shape of spacetime, the curving of spacetime, to cause this effect of gravity, one could propose an equivalent shape change on a *non-point sized particle*. The change in the shape of spacetime in the local region of the particle is mirrored by an identical change in the shape of the particle. This is not macro body deformation due to the gravitational gradient[ii] but particle-level deformation due to space contraction, time dilation and mass increase. The resulting deformation of the particle's shape is evidenced as a shift in the center of fields of its mass-volume 'field'.

This is a logical extension of both, the inertia Lorentz-Fitzgerald transformations $\Gamma(v)$, equation (7), and the Newtonian non-inertia gravitational transformation $\Gamma(a)$, equation (8),

$$\Gamma(v) = 1/\sqrt{(1 - v^2/c^2)} = x_0/x_v = t_v/t_0 = m_v/m_0 \tag{7}$$

$$\Gamma(a) = 1/\sqrt{(1 - 2GM/rc^2)} = x_0/x_a = t_a/t_0 = m_a/m_0 \tag{8}$$

Or in the generic form, the environmental transformation $\Gamma(e)$,

$$\Gamma(e) = x_0/x_e = t_e/t_0 = m_e/m_0 \tag{9}$$

Solomon [6] proposed that this mass-volume field deformation was due to the *internal* effects of the Newtonian non-inertia transformations $\Gamma(a)$, present in the local region of the *external* gravitational field, such that the spacetime transformations $\Gamma_{s(x,y,z,t)}$ are concurrently reflected as particle transformations $\Gamma_{p(x,y,z,t)}$ or,

$$\Gamma_{p(x,y,z,t)} = \Gamma_{s(x,y,z,t)}$$

(10)

The utility of equation (10) is that it explains why the gravitational field passes through all matter. Further, using the Newtonian non-inertia transformations $\Gamma(a)$ one can now replace the right hand side separation vector function in equation (3) with mass and volume spatial gradients per equation (11).

$$g = f(\Gamma(a)_{mass}, \Gamma(a)_{volume})$$

(11)

Solomon [1] proposed the new Quark Velocity model on how 'orbiting' quarks form protons and neutrons where all up & down quarks have the same velocity $v_q = 299,777,413$ m/s, in spite of different radii of 'orbits'. It is this 'orbital' velocity that gives up & down quark rest masses of 3.343167638 and 3.356973524 MeV, respectively.

In this paper, within the context of this source-field or matter-gravity schema, this Quark Velocity model is used to address equation (6), how does in the internal structure of the nuclei and atomic shells (source/matter) cause gravitational fields? The nuclei structure is modeled as circular orbits of quarks, present in the last proton or neutron added to the nuclei. The results[iii] are confirmed across 703 isotope data provided in [13], and as many as 798 isotope data provided in [14] for related model tests where empirical data is available.

3.3 Matter-Gravity Relationship

This section details the approach used and proposes how matter creates gravitational fields. A matter-gravity hypothesis has to address several issues:

i) **What is the Matter-Gravity Relationship?:** Equation (6) shows that even though our theories on gravity are very sophisticated they do not address how matter creates the gravitational field.

ii) **Unknown Systematic Error:** "Despite the increasing precision of some 300 modern-day near field experiments different labs have found slightly different values for the gravitational constant G, and in recent years the discrepancy has widened rather than narrowed" [15] & [16]. In the opinion of the author an unknown systematic error is present, therefore, a matter-gravity hypothesis would need to propose how these experiments could be improved.

iii) **Gravitational Constant G is a Variable:** A second cause of measurement discrepancy could be that the gravitational constant G is not a constant. G could be a variable dependent upon some independent *matter* related factor, thus causing systematic error. Therefore, a matter -gravity hypothesis would need to address whether the gravitational constant is a constant or not.

iv) **Mass is a Proxy for Matter:** Solomon [2] had proposed that mass was a proxy for matter. The G measurement discrepancies would affirm such a proposal as the same amount of mass does not deliver repeatable measurements. Therefore, if such a proxy relationship does exist, any matter-gravity hypothesis will need to address, how mass could be a proxy for matter.

v) **Mass of Heavenly Bodies:** Because the Earth is our laboratory, the mass of any heavenly body is determined by the product of the Earth's mass, M_E, and the Earth-based gravitational constant, G_E, or $G_E M_E$, and therefore, even though $G_E M_E$ and $G_H M_H$ for heavenly body H, is well understood, variations in G_E will cause variations in known masses M_H of these heavenly bodies. Should the gravitational constant G, change by some unknown factor, would it alter the estimated masses of the heavenly bodies?

3.4 Determining Independent Factors

The empirical data [13] & [14] was used to determine possible matter-gravity relationships, and numerical models were built to test these relationships. A good clean relationship would require an R^2 greater than 99.9999%[iv].

$$g = GM/r^2 \qquad (12)$$

$$G_i = (r_i c^2 / 2M_i)(1 - 1/t_i^2) \qquad (13)$$

For example, equation (13), derived from equations (8) & (12) was the initial prototype relationship for isotope i's gravitational constant G_i, of mass M_i, experiencing a time dilation t_i, and at distance r_i. Subsequently different relationships were tried for better fit. This was an iterative process until relationships with R^2 greater than 99.9999% were determined.

Equation (1) shows that time dilation and distances are independent factors on the field side of a matter-gravity hypothesis.

On the matter side, equation (12) shows that mass is an independent factor, but with G measurement discrepancies, one had to dig deeper. Could the number of protons & neutrons or up & down quarks be a better determinant of G? Does the nucleus' empirical RMS charge radius $r_{i,c}$ and/or empirical atomic shell radius $r_{i,a}$ also contribute to G?

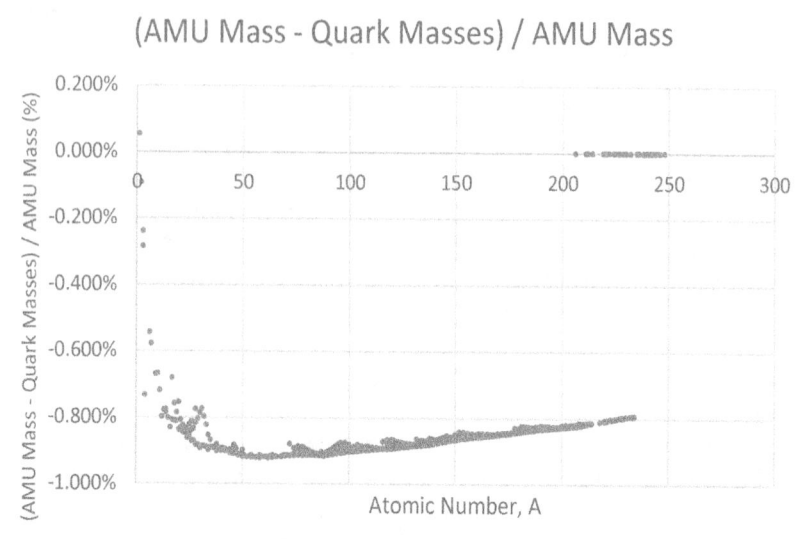

Fig. 1, *Deviation from AMU mass, (AMU – Quark)/AMU masses*

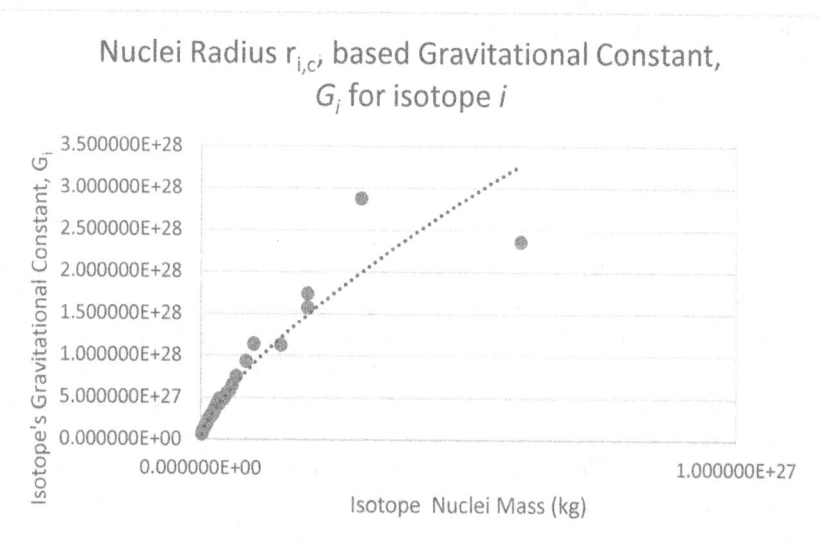

Fig. 2, *Isotope's gravitational constant, Gi, given the nucleus' RMS charge radius.*

There are two measures of mass for the atomic nuclei, i) sum of quark masses = (#up quarks x mass of up quarks) + (#down quarks x mass of down quarks); and ii) AMU mass = (nuclei AMU) x (mass of 1 AMU). Fig. 1 depicts the difference between the two approaches.

Fig. 1 shows that the differences between AMU mass and quark masses is non-linear and primarily due to binding energy. Therefore, quark mass was selected as the independent factor for matter, because it represents the total mass/energy of the nuclei and with the Quark Velocity [1] model it's definition is much cleaner and simpler than nuclei AMU.

In the first evaluation, the radius r_i is the nuclei's RMS charge radius $r_{i,c}$ of isotope i, and t_i is the time dilation at the radius. Using the Quark Velocity model of 299,777,413 m/s, time dilation $t_Q = 93.422806$.

Fig. 2 shows that the isotope's gravitational constant G_i (for 798 isotopes) is not a clean function of the isotope mass M_i given the Quark Velocity model. Thereby eliminating the RMS charge radius $r_{i,c}$ as an independent factor, for this specific model. The Quark Velocity model requires constant quark 'orbital' velocities, and thus equation (13) reduces to equation (14), where k is some constant.

$$G_i = (r_{i,c}/M_i)k \tag{14}$$

Using equation (13) these calculation were repeated with the empirical atomic radius for $r_{i,a}$. After several tweaks, rewriting equation (1) as (15) provided the best fit, Fig. 3, and affirming the Quark Velocity model.

$$(t_Q - t_{i,a}/r_{i,C} - r_{i,a})c^2 = k_s/r_{i,a} + k_c \tag{15}$$

Where, t_Q is quark velocity time dilation at the RMS charge radius $r_{i,c}$ of isotope i, and $t_{i,a}$ the gravitational time dilation at the atomic shell radius $r_{i,a}$; *slope constant k_s = -8.306600x10^{18}$ and *intercept constant k_c = - 1.674976x10^{22}$, with $R^2 > 99.999999\%$.*

3.5 Isotopic Gravitational Constant

Having determined that the empirical atomic radius $r_{i,a}$, of the 703 isotopes, yielded the relationship as depicted in Fig. 3, a new search with the new results was conducted. A new relationship, equation (16), was discovered between the isotopic gravitational constant G_i and the mass of the isotope M_i (kg).

$$G_i M_i = k_{iso} \tag{16}$$

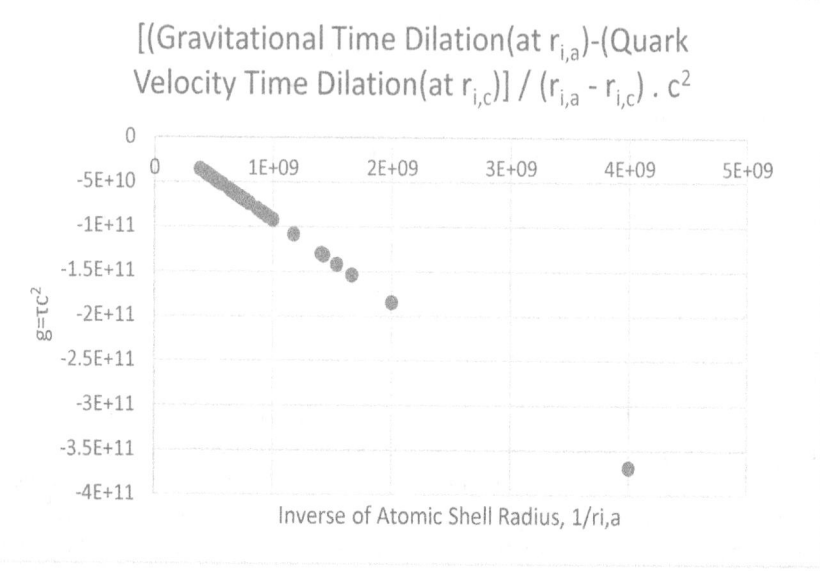

Fig. 3, Isotope's gravitational constant, Gi using the empirical atomic shell radius.

Where *isotope constant*, k_{iso} = 2.973856x10^{-36} m³s⁻². Thereby, proving that the gravitational constant G, is not a constant but a variable G_i, dependent on the isotopic mass M_i.

For non-trivial gravitational bodies, the gravitational acceleration g_H of a heavenly body of mass M_H at a distance R_H is given by its weighted w_i isotopic composition of the respective isotopic gravitational constants G_i,

$$g_H = k_{a,R}\left(\sum_i w_i G_i\right) M_H/R_H^2$$

(17)

Where the *aggregation constant at radius* R, $k_{a,R}$ = 2.244171x10²⁵ for Earth-based observations, such that,

$$k_{iso}k_{a,R} = G$$

(18)

That is, the well-known gravitational constant G = 6.67384x10⁻¹¹ *m³kg⁻¹s⁻²*.

Rearranging equation (13) the time dilation t_i, caused by a single isotope nucleus, at a distance r_i, is given by,

$$t_i = 1/\sqrt{1-(2G_iM_i/r_ic^2)}$$

(19)

or

$$t_i = 1/\sqrt{1-(2k_{iso}/r_ic^2)}$$

(20)

Given the Quark Velocity model, 'orbiting' quarks (v_q = 299,775,283m/s) inside the nucleus, alter spacetime outside the nucleus, such that the transformations present in spacetime obey equations (19) & (8) to produce gravitational fields.

In effect at the boundary of the RMS charge radius $r_{i,c}$, there is an *impulse type step-down* from the tangential quark velocity v_q, to the *radial step-down* equivalent to the gravitational escape velocity $v_{s,r}$, shown in Fig 4. This implies that the nuclei matter medium^v is substantially different from that of spacetime, and has special properties. The empirical *impulse type step-down* velocity relationship is,

$$v_{s,r}^2 \, r_{i,c} = 2k_{iso} \tag{21}$$

Equation (21) therefore, determines the time dilation present at the outer boundary (RMS charge radius $r_{i,c}$) of the nuclei. At this boundary these transformations, equation (8), present at the starting of the gravitational field spread out from $r_i = r_{i,c}$ into spacetime, to $r_i = \infty$, as a function of the radial distance r_i.

In gravitational fields, the relationship between orbital (tangential) v_{orbit} and escape (radial) velocity v_{escape} is

$$v_{orbit} = v_{escape} / \sqrt{2} \tag{22}$$

Therefore the equivalent *step-down* tangential velocity $v_{s,t}$ is,

$$v_{s,t}^2 \, r_{i,c} = k_{iso} \tag{23}$$

Just as light refracts in slower materials, the *impulse type step-down function* could be due to nuclear refraction n, that the quark velocity v_q refracts to the step-down tangential velocity $v_{s,t}$,

$$n = v_q / v_{s,t} = v_q \sqrt{r_{i,c} / k_{iso}} \tag{24}$$

$$n = \left(v_q \sqrt{1/k_{iso}} \right) \sqrt{r_{i,c}} = k_n \sqrt{r_{i,c}} \tag{25}$$

Where *nuclear refractive constant* $k_n = 1.73834 \times 10^{26}$. This is how quark motion generates gravitational fields.

3.6 Consequences of a Variable G

i) **What is the Matter-Gravity Relationship?:** Equations (21),(23) & (25) determine how matter creates the gravitational field. By altering spacetime at the RMS charge radius $r_{i,c}$ and the *step-down* velocity $v_{s,r}$, spreads radially throughout spacetime, from $r_i = r_{i,c}$ to $r_i = \infty$, as gov-

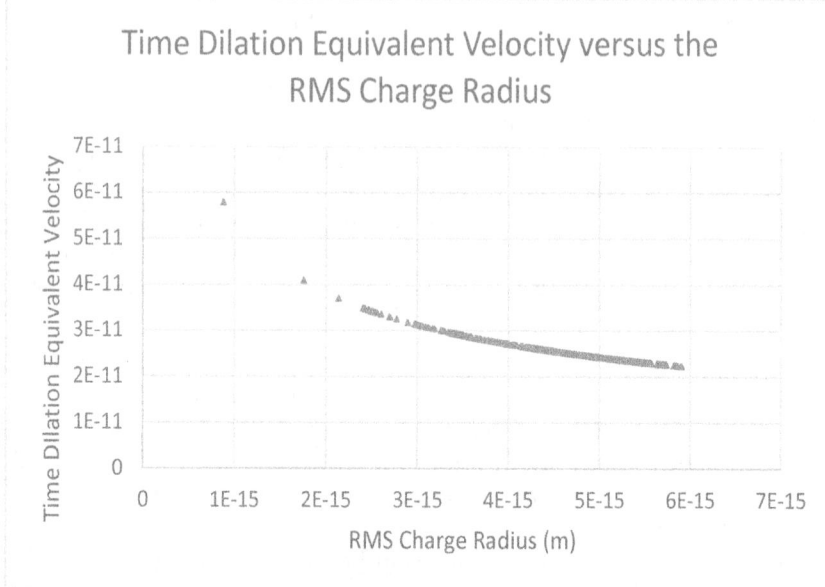

Fig. 4, Gravitational time dilation equivalent (escape) velocity at the RMS charge radius boundary of the nuclide and spacetime.

erned by equation (8). This velocity $v_{s,r}$ is a measure and determinant of the transformations present in spacetime.

ii) ***Unknown Systematic Error:*** One can now propose that systematic errors are present in all near field gravitational experiments. This is due to variations in the isotopic gravitational constant G_i, per the material used to cause variations in the gravitational force. Specifically, when $G_i > G_E$ the variations in gravitational force will be greater and vice versa. It would also be necessary to predetermine (a) the composite gravitational constant G_C, of the surrounding location due to local differences in the isotopic composition at that part of the Earth's crust, and (b) the test mass as opposed test weight.

iii) ***Gravitational Constant G is a Variable:*** The gravitational constant G, is correctly a variable G_i, equation (17). The hyperbolic relationship, equation (16), between G_i & M_i explains why Earth based gravitational constant G_E has only been observed as a constant.

iv) **Mass is a Proxy for Matter:** Note that, for a single nucleus i, the gravitational mass M_H is the mass of the nucleus M_i, and equation (16) reduces to,

$$g_i = k_{a,R}G_iM_i/R_i^2 = k_{a,R}k_{iso}/R_i^2 = G_E/R_i^2 \qquad (26)$$

That is, for a single atom, its gravitational field acceleration is independent of the mass of the atom source. Therefore, mass is a proxy for the amount of matter. Checking equation (26) with (17) shows that the aggregation constant term is not constant over radial distances at subatomic sizes, shown in Fig. 5 and equation[vi] (27),

$$k_{a,R} = k_sR_{i,C}^{k_e} \qquad (27)$$

Where the *slope constant* k_s=7.425421x10^{-26} and the *exponent constant* k_e=-3.477708.

Therefore, when aggregating mass into a gravitating body, it is necessary to know specifically,

i)   the amount (due to clustering of matter, M_H),

ii) the type (due to G_i), and

iii) the arrangement (due to $k_{a,R}$) of matter causing the gravitational field. In effect equation (17) can be written as

$$g_H = (k_{a,R}M_H)\left(\sum_i w_iG_i\right)/R_H^2 \qquad (28)$$

With $k_{a,R}M_H$ the aggregating factor and Σw_iG_i the composite gravitational constant, G_C.

i)   **Mass of Heavenly Bodies:** Equation (15) shows that the measured mass of the heavenly bodies will be dependent upon the isotopic composition of their matter.

For example, Earth composition[vii] of 35.0% Fe, 30.0% O, 15.0% Si, 13.0% Mg, 2.4% Ni, 1.9% S, 1.1% Ca, 1.15% Al, and 0.5% others, gives a weighted average gravitational constant of 6.673840x10^{-11}.

One can determine Jupiter's mass is 22.6 times less at 8.391321x10^{+25}kg instead of the current estimate of 1.900x10^{+27}kg. This was arrived at by using i) the value of the *aggregation constant* $k_{a,R}$ provided in equation (17); ii) isotopic gravitational constants of Hydrogen (^{1}H) and Helium (^{4}He) of 1.777957x10^{-09} and 4.441839x10^{-10}, respectively; iii) Jupiter[viii] consists of 80% hydrogen and 20% helium; iv) giving a composite gravitational constant $G_J = 1.511120x10^{-09}$ for Jupiter.

Super Physics for Super Technologies

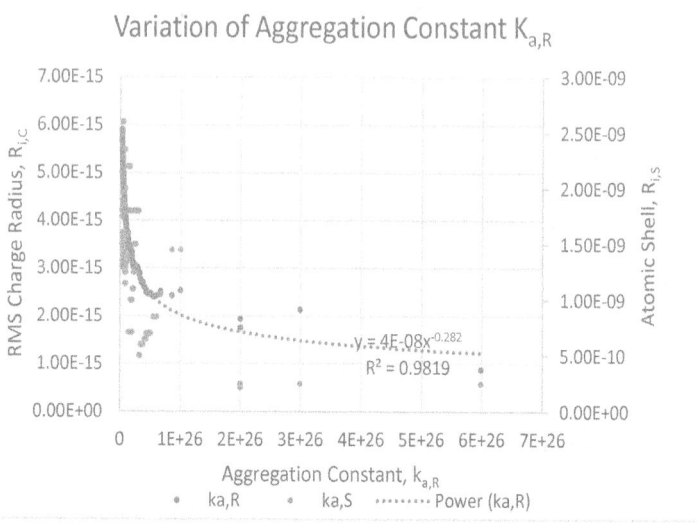

Fig. 5, *Radial Variation of Aggregation Constant Term* $k_{a,R}$.

Table 1: Universe's Composite Gravitational Constant G_U of as the composition of matter changes

Big Bang Estimates	Type	Mass (kg)	Composition	G_i	Composite G_U	Change (%)
1 second after Big Bang	Protons	1.672625E-27	99.99%	1.777957E-09	1.777957E-09	
	Neutrons	1.674927E-27	0.01%	1.775513E-09		
Primor-dial	Hydrogen	1.672625E-27	99.98%	1.777957E-09	1.777779E-09	0.010005%
	Deuteron	3.347550E-27	0.02%	8.883678E-10		
First Few Minutes	Hydrogen	1.672625E-27	75.00%	1.777957E-09	1.444514E-09	18.754282%
	Helium	6.695100E-27	25.00%	4.441839E-10		1.2x smaller
Current Day					6.673840E-11	96.246343% 26.6x smaller

Thus, the determination of the true mass of a heavenly body will be dependent upon its isotopic composition.

ii) **Unintended Consequences:** As a result of the variations in the composite gravitational constant G_C $(=\Sigma w_i G_i)$ of a heavenly body, one can expect a star's G_C to decrease as nucleosynthesis converts Hydrogen into heavier more complex nuclei. This reduced G_C will reduce the gravitational pressures and thereby reduce this conversion rate.

iii) **Cosmological Inflation:** It is suggested that nucleosynthesis caused cosmological inflation[ix] in the early Universe. The nucleosynthesis[x] of Hydrogen (${}^1H$ $G_i = 1.777957 \times 10^{-09}$) to lighter elements, Deuterium (${}^2H$ $G_i = 8.883678 \times 10^{-10}$), Helium (${}^4He$ $G_i = 4.441839 \times 10^{-10}$) and

Lithium (^{6}Li G_i = 2.961226 x 10^{-10} and ^{7}Li G_i = 2.537945 x 10^{-10}) caused a reduction in the Universe's composite gravitational constant G_U. This in turn caused changes in the clustering of matter and the reduction of the *aggregation constant at radius R*, $k_{a,R}$.

iv) **Big Bang:** Similarly, it is proposed that the composite gravitational constant G_U of the Universe, decrease more than 26.64x from a value greater than ^{1}H's G_i = 1.777957 x 10^{-09}, to the current Earth-based G_E = 6.673840 x 10^{-11}. This reduction was due to the creation of heavier, more complex particles, enabling entropy to expand the Universe[xi] at the rate at which the Universe's composite gravitational constant G_U decreased. Table 1, shows that the Universe's composite gravitational constant G_U reduced by 18.8% within the first few minutes of the Big Bang.

v) **Galaxy Rotation Curves Anomaly:** The constant rotation[xii] of a galaxy's outer stars could be due to the galaxy's composite gravitational constant G_C decreasing outwards. That as a star ages its G_C decreases, causing its galactic orbital velocity to reduce while moving the star outwards. This suggests a radial age gradient with older stars further away from the center of the galaxy and younger stars closer to the center.

Brownstein & Moffat [17] had suggested that the gravitational constant could be modified as,

$$G_\infty = G_0 \left(1 + \sqrt{M_0/M}\right)$$

(29)

Where G_∞ is the effective gravitational constant at infinity, G_0 the "bare" Newton's gravitational constant, M the mass of the galaxy and M_0 a coupling constant. This is interesting as it modifies the gravitational constant at infinity from the matter source. By comparison equation (17) modifies the composite gravitational constant G_C of local matter in local spacetime by,

$$G_C = \sum_i w_i G_i$$

(30)

One would expect G_C to be a function of the stars' ages in the galaxy, and therefore a "historical" record of the galaxy's formation. Examining Brownstein & Moffat [17] 101 galaxies for their gravitational rotational curves provides some surprises.

The three examples provided here are illustrations of these curves:

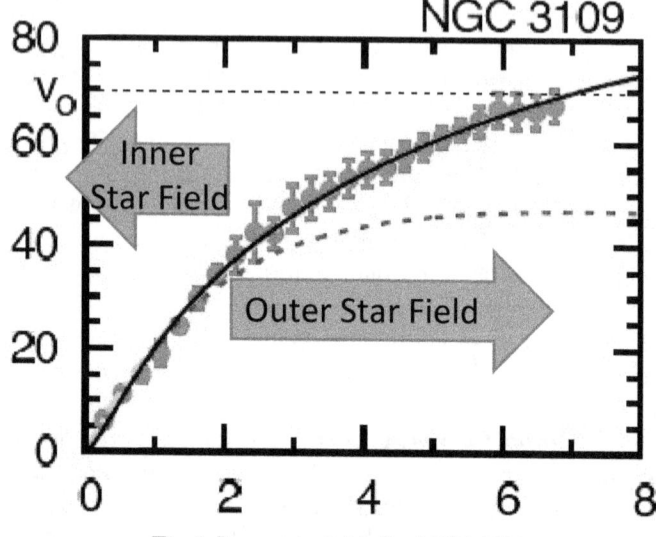

Fig. 6, Brownstein & Moffat, NGC 3109

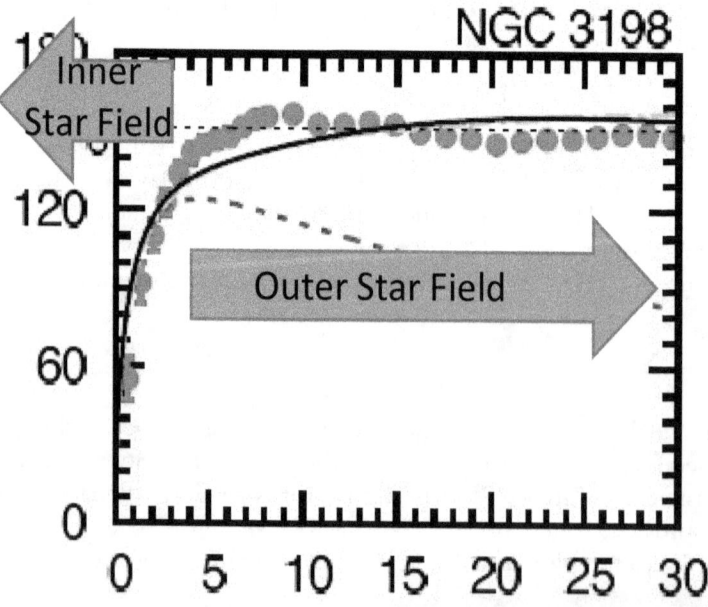

Fig. 7, Brownstein & Moffat, NGC 3198

i) Fig. 6, NGC 3109 which has a continuous positive gradient curved slope with no flat region.

ii) Fig. 7, NGC 3198 which after a good straight slope, changes to flat rotational velocities.

iii) Fig. 8, NGC 3034 has a negative slope.

To derive useful conclusions, four steps were taken to structure these apparently random collection of rotational curves. The axis sequence proposed, Table 2, is based on the assumption of consistent galactic processes.

If in early galaxy formation the bulk of the galactic mass was essentially at the center (lumps), one would expect a negative spatial velocity gradient for the Inner Star Field (ISF). As this stellar mass spreads out, this spatial velocity gradient would change to the observable *steep*, then *intermediate* and finally *gentle*. This recession of ISF stars to the Outer Star Field (OSF) is due to the conversion of hydrogen ($G_i = 1.777957 \times 10^{-09}$) to helium ($G_i = 4.441839 \times 10^{-10}$) causing the star's composite gravitational constant G_C to drop substantially by 18.8% using current ^{1}H:^{4}He ratios. If this stellar birthing galactic mass is still very much in the center, then the OSF spatial velocity gradient would be *negative*. As, this galactic mass converts to stellar mass, it spreads out (due to decreasing G_C) and the spatial velocity gradient becomes *flat*. With further mass spreading, the spatial velocity gradient becomes *positive*.

For outward recession,

$$R_A > R_B \tag{31}$$

The relationship between a star's orbiting radius R_B, *before* significant ^{1}H:^{4}He conversion, and the orbiting radius R_A, *after* significant ^{1}H:^{4}He conversion, can be shown to be,

$$V_A^2 < (G_A/G_B)V_B^2 = k_{H:He}V_B^2 \tag{32}$$

Where G_B and G_A are the composite gravitational constants, and V_B and V_A are the stellar orbiting velocities, *before* and *after* the ^{1}H:^{4}He conversion, respectively. By conservation of energy, for a star of mass m, and central galactic mass M, the total kinetic and potential energy, before T_B and after T_A must be the same,

$$T_B = mV_B^2/2 + G_B Mm/R_B \tag{33}$$

$$T_A = mV_A^2/2 + G_A Mm/R_A \tag{34}$$

Using equation (31),

$$T_A = mk_{H:He}V_B^2/2 + k_{H:He}G_B Mm/R_A \tag{35}$$

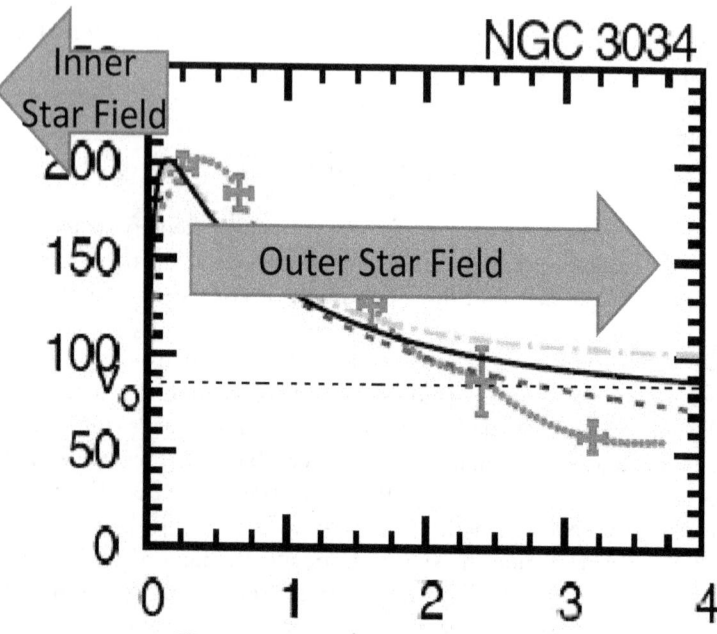

Fig. 8, Brownstein & Moffat, NGC 3034

Table 2(a): Galactic Stages (#)

Outer Stars		Negative	Flat	Positive	Totals
Inner	Steep	23	23	1	48
Stars	Intermediate	2	12	5	19
	Gentle	2	6	26	34
	Totals	27	41	33	101

Table 2(b): Galactic Stages (%)

Outer Stars		Negative	Flat	Positive	Totals
Inner	Steep	22.8%	22.8%	2.0%	47.5%
Stars	Intermediate	2.0%	11.9%	5.0%	18.8%
	Gentle	2.0%	5.9%	25.7%	33.7%
	Totals	26.7%	40.6%	32.7%	100.0%

Therefore,
$$T_B > T_A \qquad (36)$$
i.e. there is a missing energy component $f(G_A)$,
$$T_B = T_A + f(G_A) \qquad (37)$$

The ^{1}H:^{4}He conversion requires $G_B > G_A$, and causes the missing energy component $f(G_A)$. This introduces the possibility that $f(G_A)$ has been interpreted as dark energy as the Universe is anchored at T_B, while our observations are at T_A.

The second step was to use this proposed structure, and visually analyze & categorize the 101 Brownstein & Moffat [17] rotation curves. Table 2, shows that only 4 galactic stages account for 83% of these galaxy structures.

Steep ISF velocity gradients are accompanied by either *negative* (22.8%) or *flat* (22.8%) OSF spatial velocity gradients. ISF stars with *intermediate* gradients are accompanied by *flat* (11.9%) OSF spatial velocity gradients. And ISF stars with *gentle* gradients are accompanied by *positive* (25.7%) OSF spatial velocity gradients.

The third step was to formalize this ISF & OSF spatial velocity gradients into matrix (Fig. 9). The sample statistics in Table 2 suggest that some galactic stages are much more likely (high counts) than others. On a galactic scale, some stages are short lived and have low counts. This provides a guide to determining how galaxies mature. The logical conceptual time line of how galaxies emerge out of dust clouds or galaxy life cycle, Fig. 9, consists of the Main Sequence and two minor sequences.

The Main Sequence is identified by the red solid arrows, and comprise of stages 1, 2, 5, & 9, accounting for 84 (83.2%) of the Brownstein & Moffat [17] rotational curve data. The two minor sequences are the 'Left Sequence' (1, 4, 5, 7, 8 & 9) depicted by the purple dotted arrows, and 'Right Sequence' (2, 3, 5, 6, & 9) depicted by the orange striped arrows. Our Milky Way is a Main Sequence, Stage 2 galaxy.

This time line would suggest that black holes at the center of galaxies are a later development formed by the capture of stars.

The fourth and final step was to use the data from the 101 rotational curves to investigate possible relationships between mass & velocities.

To extract sufficiently good data, the 101 rotational curves were structured as shown in Fig 10. These curves consist of the two ISF & OSF spatial velocity gradients (solid red lines) and three regions of noisy data (purple boxes).

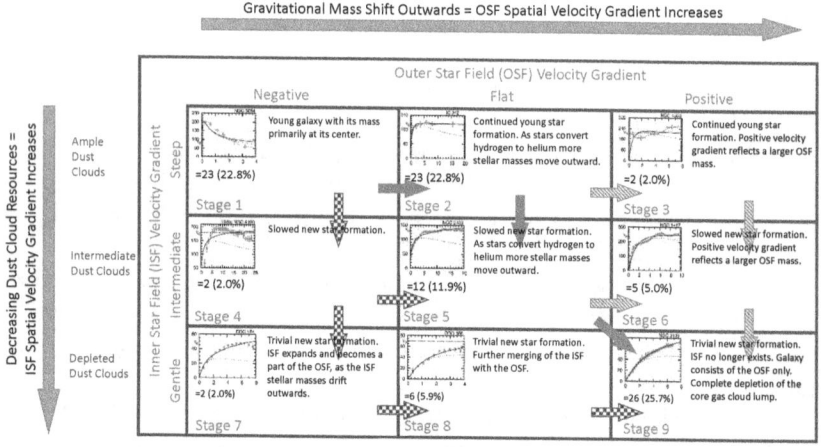

Fig. 9, Galactic Main Sequence

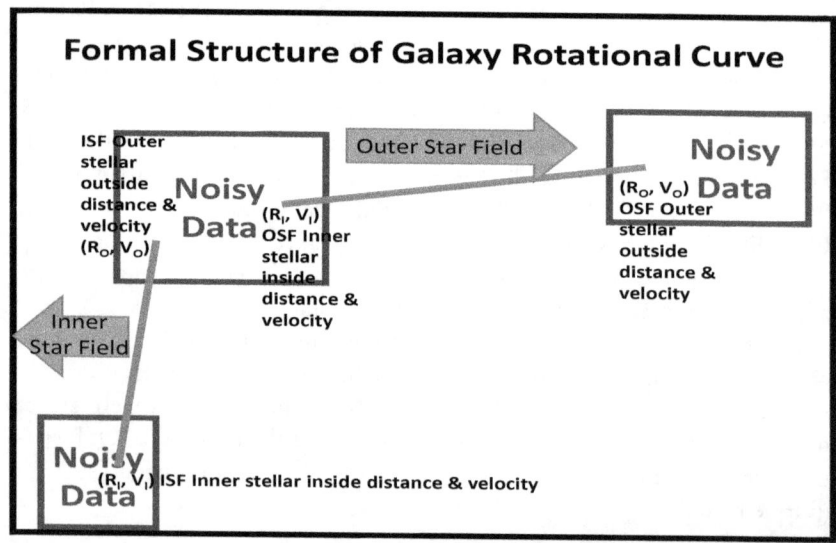

Fig. 10, Formalized Rotational Curve Structure

Using this structure, it was found by visual inspection, only 35 of the 101 rotation curves provided good spatial velocity gradient data.

This 35 were then regression *(74.7% < R² < 99.1%)* fitted to determine a 'best' type of relationship, given the logic that if there were a consistent galactic process:

i)   This consistent galactic process would be revealed as a consistent type of equation, linear, exponential, logarithmic, power or polynomial. The polynomial is the least desired as most fundamental physics are not polynomials.

ii) Variations in this consistent galactic process would be evident as variations in the coefficients of this equation.

Assuming constant galaxy mass M, and substituting centripetal acceleration into equation (17) and dividing outer and inner star equations (see Fig 10) gives equation (38) for both ISF & OSF, while M cancels out. For ease of usage let's name the LHS as the *ratio function*.

$$\frac{k_{a,O}\left(\sum_i w_i G_{i,O}\right)}{k_{a,I}\left(\sum_i w_i G_{i,I}\right)} = \frac{V_O^2 R_O}{V_I^2 R_I} \tag{38}$$

Where $k_{a,O}$ & $k_{a,I}$ are the aggregation constants and $G_{i,O}$ & $G_{i,I}$ the isotopic gravitational constants of the outer and inner stars, respectively. The best fit for the RHS of equation (38), are shown in Fig. 11, Fig. 12 and Fig. 13.

This analysis shows that there are only two types of *ratio functions* and therefore, processes, at work. The exponential process,

$$\frac{k_{a,O}\left(\sum_i w_i G_{i,O}\right)}{k_{a,I}\left(\sum_i w_i G_{i,I}\right)} = p e^{q\frac{(R_O - R_I)}{R_O}} \tag{39}$$

Where p and q are constants such that,

i)   For OSF *positive* & *negative* gradients,

$$0.262448 < p < 0.600561 \tag{40}$$
$$3.048703 < q < 3.878342 \tag{41}$$

ii) For all ISF gradients,

$$4.268896 \times 10^{-6} < p < 1.050642 \times 10^{-3} \tag{42}$$
$$13.934100 < q < 19.019377 \tag{43}$$

However, the OSF *negative* spatial velocity gradient ratio function is a linear function, not exponential,

$$\frac{k_{a,O}\left(\sum_i w_i G_{i,O}\right)}{k_{a,I}\left(\sum_i w_i G_{i,I}\right)} = p\left(\frac{V_O}{V_I}\right) + q \tag{44}$$

Where

$$-9.930104 < q < -1.256407 \tag{45}$$

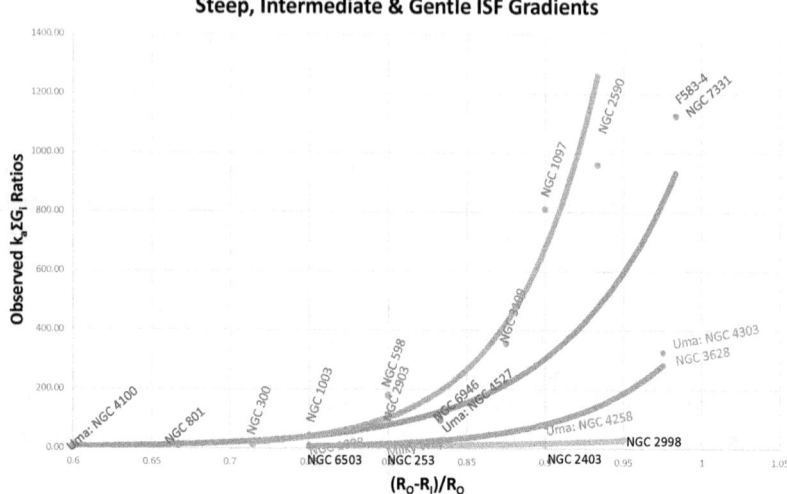

Fig. 11, Observed $k_a\Sigma G_i$ Ratios

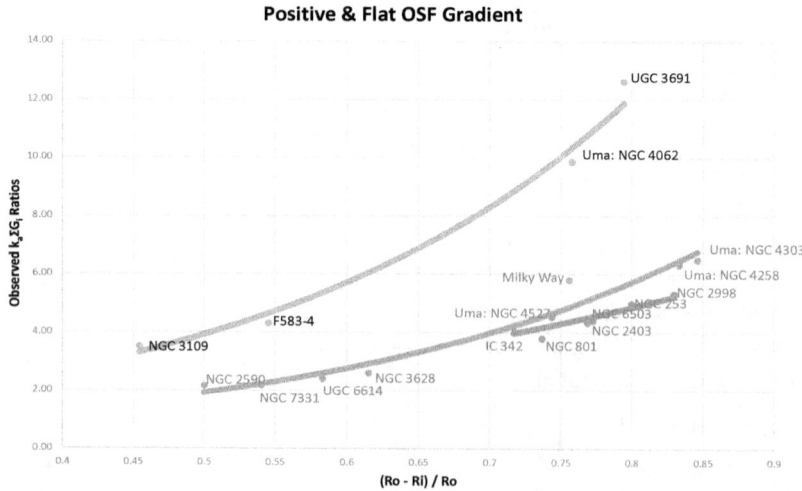

Fig. 12, Observed $k_a\Sigma G_i$ Ratios

$$3.913963 < p < 17.199780 \tag{46}$$

That, the ISF and OSF formation processes are similar, except for the negative OSF gradient process.

Simplifying equation (38) with (47) gives (48)

$$\sum_i w_i G_{i,O} \approx \sum_i w_i G_{i,I} \tag{47}$$

$$\frac{k_{a,O}}{k_{a,I}} \approx p e^{q \frac{(R_O - R_I)}{R_O}} \tag{48}$$

Equation (48) shows that the ratio of aggregation constants $k_{a,O} / k_{a,I}$ is a function of the range R_O - R_I, and size R_O of the galaxy. Or given a consistent galactic process, the distribution of matter across the galaxy, plays a critical role in the nature of the rotational curves. It appears that $k_{a,R}$ undergoes extreme changes at subatomic distances and very much smaller changes at galactic distances.

This analysis had four surprises. First, that there is a term $k_{a,R}$ for the arrangement of matter as opposed to the quantity of matter. Second, the existence of ISF & OSF necessarily requires that the aggregation of matter $k_{a,R}$, is different between the two, suggesting a lumpy origin. Third, the negative spatial velocity gradients are of a different process type to the other galactic processes, linear V_O / V_I, versus exponential $(R_O\text{-}R_I)/R_O$, respectively. And fourth, NGC 3034's OSF *ratio function* < 1.

$$\frac{k_{a,O}\left(\sum_i w_i G_{i,O}\right)}{k_{a,I}\left(\sum_i w_i G_{i,I}\right)} < 1 \tag{49}$$

3.7 Conclusion

This paper has shown that matter, not mass, creates gravitational fields and mass is a proxy for matter. That, the gravitational constant G is not a constant and is replaced two factors. First, the *aggregation constant at radius R*, $k_{a,R}$, and second, the *isotopic gravitational constant G_i*. The isotopic gravitational constants result in new ways of looking at the Universe as a function of the reducing composite gravitational constant G_C. Thus, the time changing gravitational constant and the proposal of a galactic Main Sequence. However, answering the question, how the increase in atomic charge decreases the *isotopic gravitational constant G_i*? will lead to a better understanding of the relationship between gravity and electric fields, and better particle models.

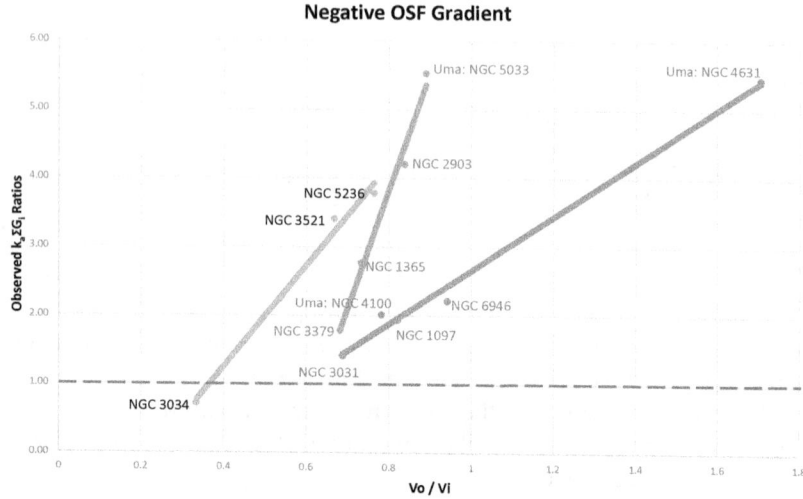

Fig. 13, Observed $k_a \Sigma G_i$ Ratios

References:

[1] Solomon, B.T., "A Universal Approach to Forces", in *Super Physics for Super Technologies*, March 2015.

[2] Solomon, B. T., "Empirical Evidence Suggest A Need For A Different Gravitational Theory," in the proceedings of the *100 Year Starship Study Public Symposium (100YSS,)*, 2013.

[3] Solomon B.T., "New Evidence, Conditions, Instruments & Experiments for Gravitational Theories", Journal of Modern Physics, Special Issue on Gravitation, Astrophysics and Cosmology, Vol. 8A, 2013, August 2013.

[4] Solomon, B. T., "Empirical Evidence Suggest A Need For A Different Gravitational Theory," *American Physical Society (APS) April Conference, Denver*, 2013

[5] Solomon B.T., *An Introduction to Gravity Modification: A guide to using Laithwaite's and Podkletnov's experiments and the physics of forces for empirical results*. Universal Publishers, Boca Raton, 2nd Edition, May 2012.

[6] Solomon, B. T., "Non-Gaussian Radiation Shielding," *100 Year Starship Study Public Symposium (100YSS,)*, 2011.

[7] Solomon, B.T., "Gravitational Acceleration Without Mass And Noninertia Fields", Physics Essays, Vol. 24, 327, 2011. [Phys. Essays **24**, 327 (2011)]

[8] Solomon, B. T., "Reverse Engineering Podkletnov's Experiments," in the proceedings of the *Space, Propulsion & Energy Sciences International Forum (SPESIF-11)*, Edited by Glen A Robertson, Physics Procedia, Elsevier Science.

[9] Solomon, B. T., "Non-Gaussian Photon Probability Distributions," in the proceedings of the *Space, Propulsion & Energy Sciences International Forum (SPESIF-10)*, Edited by Glen A Robertson, AIP Conference Proceedings **1208**, Melville, New York, (2010).

[10] Solomon, B. T., "An Approach to Gravity Modification as a Propulsion Technology," in the proceedings of the *Space, Propulsion & Energy Sciences International Forum (SPESIF-09)*, Edited by Glen A Robertson, AIP Conference Proceedings **1103**, Melville, New York, (2009).

[11] Solomon, B. T., "An Epiphany On Gravity", Journal of Theoretics, Vol. 3-6, 2001.

[12] Feynman, R., *Feynman Lectures on Physics*, Chapter 28-4.

[13] C.W. Misner, K.S. Thorne, J.A. Wheeler, *Gravitation* (W.H. Freeman and Company, New York, NY, 1973).

[14] Audi, G., Wapstra, A.H., Nuclear Physics A595 vol. 4 p.409-480, December 25, 1995 (Excel Spreadsheet prepared by Dr. Gordon Gilmore of Nuclear Training Services Ltd.)

[15] Angeli, I., "A consistent set of nuclear rms charge radii: properties of the radius surface (R,N)", Science Direct, Volume 87, Issue 2, July 2004, Pages 185–206, DOI: 10.1016/j.adt.2004.04.002

[16] Cowen, R. "Quantum method closes in on gravitational constant. Cold rubidium atoms provide fresh approach to measuring Newton's big G", Nature, June 18 2014, doi:10.1038/nature.2014.15427

[17] Reich, E. S.,"G-whizzes disagree over gravity. Recent measurements of gravitational constant increase uncertainty over accepted value", Nature 466, 1030 (2010), doi:10.1038/4661030a

[18] Brownstein, J. R. and Moffat, J. W., Galaxy Rotation Curves Without Non-Baryonic Dark Matter, arXiv:astro-ph/0506370v4, 22 Sep 2005

End Notes -

i. Does not include non-peer reviewed conference presentations between 2001 and 2009.

ii. Macro body elongation due to tidal gravity is attributed to Roger Penrose (this author could not find the reference in time for this paper). Macro bodies elongate as the body falls into a gravitational field. Let's reexamine this tidal behavior with the additional requirement that this tidal gravity property be consistent with Lorentz-FitzGerald transformations or Special Theory of Relativity. To be consistent with Lorentz-FitzGerald transformations, the atoms and elementary particles would contract in the direction of the fall. However, to be consistent with tidal gravity's elongation, the distances between atoms in the macro body has to increase at a rate consistent with the acceleration and velocities experienced by the various parts of the macro body. That is, as the atoms get flatter, the distances apart get longer. One suspects that this axiom's inconsistency with the empirical evidence has led to an explosion of string theories, each trying to explain Nature with no joy.

iii. Note all calculation were evaluated to 250 significant digits in MS Excel using the Add In, XNumbers.

iv. Equation (15) can be resolved into a 3rd or 5th degree polynomial in $r_{i,a}$ (if k_C is set to 0 or $-1.674976x10^{22}$) and used to determine $r_{i,a}$.

v. Per [2] there is a need to propose alternatives to Relativity, String & Quantum (RSQ) theoretical approaches.

vi. $R^2 = 98.187\%$

vii. Case Study: The Natural Abundance of Elements, UC Davis ChemWiki, http://chemwiki.ucdavis.edu/Inorganic_Chemistry/Case_Studies/Case_Study%3A_The_Natural_Abundance_of_Elements

viii. Jupiter, the largest planet, Astronomy Today, http://www.astronomytoday.com/astronomy/jupiter.html

ix. http://www.ctc.cam.ac.uk/outreach/origins/inflation_zero.php

x. http://astro.berkeley.edu/~mwhite/darkmatter/bbn.html

xi. http://www.ctc.cam.ac.uk/outreach/origins/big_bang_one.php

xii. http://astro.berkeley.edu/~mwhite/darkmatter/rotcurve.html

xiii. http://stardate.org/astro-guide/btss/galaxies/galaxy_formation

xiv. From 100% ^1H to 75% ^1H & 25% ^4He, G_C would decrease 18.8% from 1.777957×10^{-09} to $1.444514x10^{-09}$.

xv. Per equation (17), the aggregation constant k_a would have a role in this analysis, but at this point would obfuscate the clarity of the discussion. See equation (47) for a fuller treatment.

4. A New Nucleon/Nuclei Structure

Abstract:- The four main issues with the *Standard Model (SM)* raise the need for an alternative theoretical approach that falsifies this Model. The four issues are i) quantum foam does not exists, ii) probabilities are not Gaussian, iii) particles are compressive, not inelastic (per quantum theory) or tensile (per string theories), and iv) the wave function contradicts *Special Theory of Relativity (STR)* [2].

This paper proposes a *Non Standard Model (NSM)* approach to nucleon/nuclei structure that is supported by the empirical data of 798 isotopes, and introduces the role of *Variable Electrical Permittivity (VEP)* matter and packing densities in nucleon/nuclei structure. *Variable Electrical Permittivity (VEP)* matter has dielectric properties that are inverse functions of the RMS charge radius $R_{i,c}$. This is the reason for a particle's mechanical and electrical stability.

Packing densities determine how tightly rigid spheres can be packed to form a volume. Packing densities suggest that nucleons cease to exist in the nucleus and are replaced by their component quarks.

It is suggested that the down quark is unstable and the source of radioactivity. Finally, it is shown that VEP matter is derived from binding energy.

This approach suggests that both weak and strong nuclear forces are falsifiable and no longer necessary, thereby providing an initial attempt at how theoretical physics could be rewritten.

4.1 Premise

This paper reports the 13th paper in the 16-year ([1] to [12]) investigation into the feasibility of gravity modification, proposing:

i) Non Inertia (Ni) Fields defined as the *spatial* gradient of real or virtual velocities.

ii) Equation (1), the massless formula, for gravitational acceleration, where τ is the change in time dilation divided by the change in that distance,

$$g = \tau c^2 \tag{1}$$

iii) All macro forces (electromagnetic, electric & magnetic, mechanical, gravitational), can be model by the spatial gradient of their fields [2].

iv) The isotopic gravitational constants that explains how matter, and not mass, causes gravitational fields.

v) The falsifiability of the Standard Model.

As a consequence of iii), that electric and magnetic field lines are not repulsive, Poincaré stresses [13] do not exist [2] thus facilitating simpler proton, neutron & atomic nuclei models.

Occam's Razor, that the simpler solution is most likely the correct solution, drives this investigation for a simpler matter model that explains how matter creates gravity.

4.2. Introduction

Starting with Nemiroff's [14] finding that quantum foam cannot exist in free space, evidence is mounting that theoretical physics is in need of a substantial rewrite. Solomon proposed [2] three important empirical based requirements of theoretical physics. (See [2] for the rest of his proposals.) These are:

i) ***Non-Gaussian Probability:*** Using the Airy disc [2], [4], [5] & [8] the empirical evidence shows that photon probability is not Gaussian but a new modified Gamma distribution, named *Var-Gamma*, equation (2), because the a & β parameters of this distribution are variable functions of the orthogonal distance r from the axis for a photon propagation of wavelength λ, and Airy disc aperture D_A.

$$(r) = \frac{1}{\beta \Gamma(\alpha)} \left(\frac{r}{\beta}\right)^{\alpha-1} e^{-r/\beta} \tag{2}$$

$$\alpha = r \tag{3}$$

$$\beta = r \sqrt{u} \tag{4}$$

$$u = \pi/(\lambda D_A) \, sin(\theta) \tag{5}$$

The mass particles' (e.g. electrons) Airy discs are consistent with photon Airy discs. Therefore, this *Var-Gamma* probability function holds for mass-based particles, and there is a consistency in particle structure design between mass & massless particles.

ii) ***Compressive Particles:*** Equation (1) was discovered [2], [4], [6] & [9] on the basis that particles are compressive and shorten in the direction of propagation/travel. A consistency with Lorentz FitzGerald transformations (LFT). This is unlike quantum or string theories [2] which require particles to be point like or expand with energy, respectively. The new finding $g = \tau c^2$ affirms that particles are compressive.

iii) ***Inconsistent Wave Function:*** All particles, with and without mass, have wave functions that spread out into the region of space surrounding the particle. In single and double slit experiments particles exhibit a consistent wave interference pattern, irrespective of whether they have mass or not. This suggests, that both photons and mass-based particles have identical wave function mechanisms irrespective of mass.

However, photons velocity v_p is the velocity of light ($v_p = c$) while mass-based particle velocity v_p is less than that ($v_p < c$). To be consistent with Lorentz-FitzGerald (LFT) and Special Theory of Relativity (STR), anything traveling at the velocity of light must have zero thickness, in the direction of propagation, and cannot spread out like the wave function does.

To be consistent with LFT & STR, the logical resolution is that the wave function is not moving in the direction of propagation/travel. The wave function's velocity $v_{wf} = 0$, is zero, and therefore, independent of v_p.

$$v_{wf} \neq f(v_p) \tag{6}$$

A zero velocity $v_{wf} = 0$ wave function is consistent with both types of particle velocities $v_p < c$ and $v_p = c$.

How could Nature implement such a property? Here is an analogy. Take a garden rake, turn it upside down and place it under a carpet. Move it. What does one observe?

The carpet exhibits a wave-function-like-envelope-bulge that appears to be moving in the direction the garden rake is moving.

But the bulge is not moving. It shows up wherever the garden rake is. The rake is moving but not the bulge. The bulge is simply an orthogonal displacement of the carpet caused by the rake.

The wave function, like the carpet bulge, is an orthogonal displacement in spacetime caused by the presence of the particle, and is not moving. The wave function spreads across the spacetime in a consistent manner whether it is a photon or a mass-based particle, and therefore, the wave-particle duality no longer exists.

This zero-velocity bulge-like wave function is now consistent with Einstein's Special Theory of Relativity (STR) and with the empirical Lorentz-FitzGerald transformations (LFT).

The *Standard Model (SM)* is successful because, just as the shape of the carpet bulge is unique to the shape of the garden rake, so is the wave function displacement in spacetime unique to the properties of the underlying particles.

4.3 Justifying New Particle Models

It was proposed [2] that the deformation of any field results in the shift in the Center of Field C_F, just as altering the shape of an object would alter its Center of Mass, C_M. The magnitude and direction of the shift in this C_F governs the strength and direction (attraction or repulsion) of the resulting motion of this Field.

Using the Center of Mass concept the Center of Field C_F of a field F that ranges from L to U, is defined as

$$C_F = \int_{L}^{U} P(x)x\,dx \left/ \int_{L}^{U} P(x)\,dx \right.$$

(7)

Where P is the property of the Field used to evaluate the Field's C_F.

The relevant field property depends upon the type of deformation applied to this field. If the deformation of the field property is non-linear (as in electric and gravitational fields) then the spatial gradient of the Field's property P or dP/dx is the parameter used to estimate the Field's C_F. If the deformation of the field property is linear (charged parallel plates) then the field's property P itself is the parameter used to estimate the Field's C_F, as the spatial gradient of P would be zero.

It was shown [2] that this Spatial Gradient Center of Field (SGCF) model does away with the need for Poincaré stresses [13] as electric field lines are not repulsive. This simplifies particle models.

Without the exchange of virtual or real particles, SGCF is an alternative universal mechanism for the transmission of all forces (gravitational, electromagnetic, electric, magnetic & mechanical).

With respect to proton & neutron structures, the Standard Model (SM) requires a "hideously complex . . . roiling infinity of quarks and gluons" [15]. The problem with infinity is that if there is even the slightest probability of an event occurring *it must*, but proton & neutron structures are stable. Therefore, suggesting that either there isn't an infinity of particles within or that probabilities cease to exists within the proton-neutron structures. The more likely possibility is that an infinity of particles don't exists because probabilities are a fundamental structure of Nature [2], [4], [5] & [8].

Therefore, a *New Standard Model* is warranted. Could this *New Standard Model* as building blocks of matter, not require particles as carriers of force? If so an alternative mechanism to holding proton-neutron structures in a stable manner is required. This paper explores the possibility of a *Non Standard Model (NSM)* particle structure with respect to proton-neutron structures.

4.4 Requires New Force Mechanism

To keep this exploration brief, this paper considers only up & down quarks as the basis for proton-neutron structures. The justification for this is the slight difference between the proton (938.272046 MeV) and neutron (939.565379) mass. Since, the proton consists of effectively 2 up and 1 down quark, while the neutron consists of effectively 1 up and 2 down quarks, the mass of these quarks must account for the difference between proton and neutron masses.

Davis *et al* [16] had suggested up and down quark rest masses of 2.01 MeV and 4.79 MeV, respectively, while Solomon [2] had suggested 3.343167638 MeV and 3.356973524 MeV, respectively. Both sets of mass estimates were arrived at by different methods. Assuming that quarks are the carrier of strong forces, with extensive simulation, Davis *et al* [16] used lattice QCD and quark mass ratios to arrive at their values.

Solomon [2] used a much simplified Quark Velocity model, that all quarks 'orbit' protons & neutrons at the RMS charge radius $R_{i,c}$ and at the same velocity v_q of 299,775,269 m/s. This approach is valid given that Poincaré stresses [13] don't exists, thus [2] electric field lines are not repulsive. Therefore, with the Quark Velocity model, quarks are no longer carriers of the strong force.

To further complicate matters, quarks have different charges; the up quark is +2/3e (electron charge) and the down quark is -1/3e. This raises

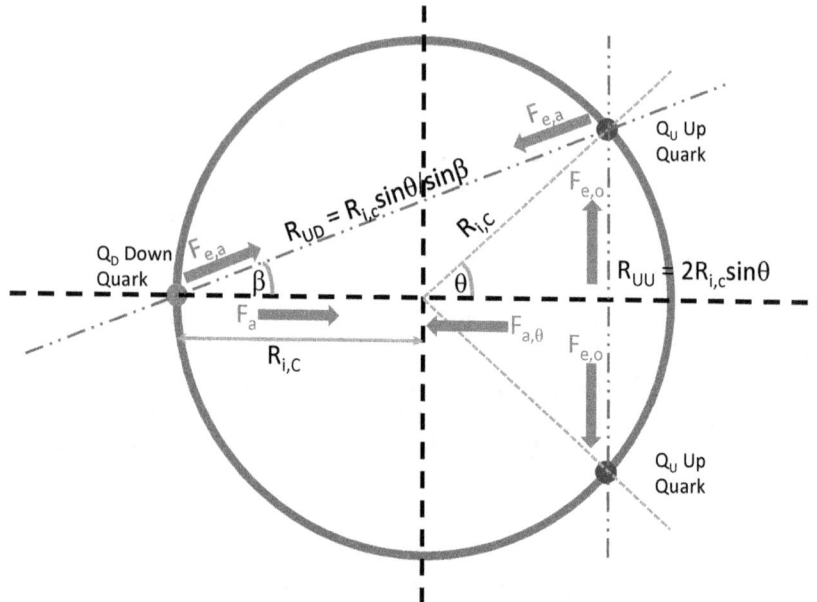

Fig. 1, Quark-Proton Model

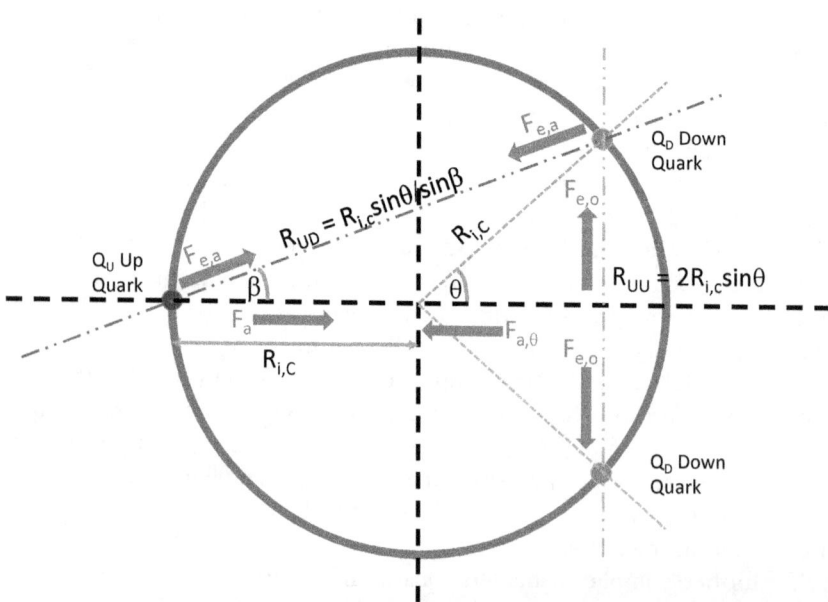

Fig. 2, Quark-Neutron Model

the question of how do 3 particles exist as a stable structure when two are attracting each other and two are repelling.

To expound how matter, and not mass, creates gravitational fields, Solomon [2] proposed that there is a substantial difference between the nuclei matter and spacetime external to it. Such that a velocity refraction, with a refraction index n, occurs at this boundary $R_{i,c}$,

$$n = v_q/v_{s,t} = k_n \sqrt{R_{i,c}} \qquad (8)$$

Where the *nuclear refractive constant* $k_n = 1.73834 \times 10^{26}$; the quark velocity v_q steps down to the gravitational field (orbital) velocity $v_{s,t}$ and starts the gravitational field at $R_{i,c}$.

The advent of negative permittivity materials, provides a potential solution. These nucleon structures enclose a nuclear matter that is *variable electric permittivity* (VEP) matter, with a permittivity ε such that, the electric permittivity of free spacetime ε_0 is altered by the *variable dielectric constant* $\varkappa$,

$$\varepsilon = \varkappa \varepsilon_0 \qquad (9)$$

This VEP matter like electron shells, alters the structure of spacetime in such a manner as to provide a *VEP Cavity (VCav)* where nucleons & nuclei exists as stable structures.

4.5 New Proton/Neutron Structure

Given high velocity quarks, the simplest model for both mechanical and electrical field stability would be concentric circular orbits at increasing RMS charge radii $R_{i,C}$ (Fig. 1 & 2) as each proton or neutron is added to the nucleus. Several assumptions & conditions are required:

i) With unknown quark shapes and sizes, their center of mass is a sufficient approximation of where they are.

ii) VEP matter responds to the presence of multiple non-repulsive [2] vector electric field lines such that $\varkappa$ is positive or negative if the electric fields lines are of the same direction (opposite sign charges) or opposite directions (same sign charges), respectively.

iii) The centripetal acceleration required to maintain the quark velocities is due to the electric field acceleration when $\varkappa$ is positive.

iv) VEP matter neutralizes the electric field acceleration when $\varkappa$ is negative.

v) VEP matter requires all particles with mass inside a nucleon or nucleus to orbit at quark velocity v_q of 299,775,269 m/s.

For mechanical stability the horizontal forces, per the orientations in Fig. 1 & 2, must cancel. The horizontal force at the single quark F_a, (masses m_U or m_D & with centripetal acceleration a_U or a_D for up & down quarks, respectively) must equal the two horizontal components of the double quarks $F_{a,\theta}$,

$$F_a = m_U a_U = 2F_{a,\theta}\cos\theta = m_D a_D 2\cos\theta \qquad (10)$$

Since, $a_U = a_D$, as all quark velocities are the same, for the Quark-Proton Model,

$$\cos\theta = m_D/2m_U \qquad (11)$$

And for the Quark-Neutron Model

$$\cos\theta = m_U/2m_D \qquad (12)$$

The electrical force of attraction $F_{e,a}$ between the up Q_U and Q_D down quarks charges is,

$$F_{e,a} = -Q_U Q_D/(4\pi\kappa\varepsilon_0 R_{UD}^2) \qquad (13)$$

R_{UD} is distance between the unlike up and down quarks and is a function of the RMS charge radius $R_{i,C}$ for isotope i. For this electrical force $F_{e,a}$ to be the origin of the centripetal force F_a, then

$$F_{e,a} = F_a \qquad (14)$$

Or the dielectric constant κ changes to ensure that the ratio of the electrical acceleration due to attraction, to the centripetal acceleration equals 1. Using regression to fit the κ relationship, equation (15), determined by equation (14) over 798 [17] isotopes gives $R^2 > 99.9999\%$ for all six regressions (16), (17), (18), (19), (23) & (24).

$$\kappa_Q = k_{c,Q}/R_{i,C} \qquad (15)$$

Where the *variable dielectric constant* κ_Q is determined by the *particle dielectric constant* $k_{c,Q}$ and the RMS charge radius $R_{i,C}$.

For the Quark-Proton Model, the dielectric constants $(R^2 = 100.000000\%)$ for the up $\kappa_{U,P}$ and down $\kappa_{D,P}$ quarks are,

$$\kappa_{U,P} = 2.644546x10^{-19}/R_{i,C} \qquad (16)$$

$$\kappa_{D,P} = 2.63364x10^{-19}/R_{i,C} \qquad (17)$$

or

$$\kappa_{U,P}/\kappa_{D,P} = M_D/M_U \qquad (18)$$

For the Quark-Neutron Model, the dielectric constants $(R^2 = 100.000000\%)$ for the up $\kappa_{U,N}$ and $\kappa_{D,N}$ down quarks are,

$$\kappa_{U,N} = 2.643151x10^{-19}/R_{i,C} \tag{19}$$

$$\kappa_{D,N} = 2.632251x10^{-19}/R_{i,C} \tag{20}$$

or

$$\kappa_{U,N}/\kappa_{D,N} = M_D/M_U \tag{21}$$

Combining all 3,192 data points into a single regression (R^2= 99.999897%), gives for attraction, up quark κ_U and down quark κ_D that are independent of the Quark-Proton or Quark-Neutron Models,

$$\kappa_U = 2.643848x10^{-19}/R_{i,C} \tag{22}$$

$$\kappa_D = 2.632945x10^{-19}/R_{i,C} \tag{23}$$

$$\kappa_U/\kappa_D = M_D/M_U \tag{24}$$

The dielectric constants are only inverse function of the RMS charge radius $R_{i,C}$ of the specific isotope i.

For electrical stability electric field repulsion needs to be neutralized. This electric force $F_{e,o}$ repulsion between like quarks, the for up and down quarks are given by,

$$F_{e,o} = -Q_U Q_U/(4\pi\kappa_u\varepsilon_0 R_{UU}^2) \tag{25}$$

$$F_{e,o} = -Q_D Q_D/(4\pi\kappa_D\varepsilon_0 R_{DD}^2) \tag{26}$$

And the radial components must equal the centripetal forces,

$$F_a = F_{e,o}\cos\theta \tag{27}$$

Thus, the respective dielectric constants (R^2= 100.000000%) $\varkappa_{2U,P}$ and $\varkappa_{2D,N}$ are,

$$\kappa_{2U,P} = -1.939188x10^{-18}/R_{i,C} \tag{28}$$

$$\kappa_{2D,N} = -4.912652x10^{-19}/R_{i,C} \tag{29)}$$

And it can be shown analytically that,

$$\kappa_{2U,P}/\kappa_{2D,N} = (Q_U^2/Q_D^2)\sqrt{(4m_D^2 - m_U^2)/(4m_U^2 - m_D^2)} \tag{30}$$

Therefore, combining these 1,596 data points (R^2=99.850467%) gives for repulsion, a consistent relationship between $\kappa_{2U,P}$ and $\kappa_{2D,N}$.

$$\kappa_{2U,P} = -1.958915x10^{-18}/R_{i,C} \tag{31}$$

$$\kappa_{2D,N} = 4.027644\kappa_{2U,P} \tag{32}$$

$$\kappa_{2D,N} = -4.863673x10^{-19}/R_{i,C} \tag{33}$$

Again, the dielectric constants are only inverse function of the isotopic RMS charge radius $R_{i,C}$.

It should be noted that as R^2>99.850% for equations (22), (23), (31) & (32) for 798 [17] isotopes or 4,788 data points, this strongly suggests that new structural relationships in nucleons include *variable electric permittivity* (VEP) matter. These properties are condensed to two relationships each, equations (22) & (23) for attraction, and equations (31) & (32) for repulsion. In other words VEP matter minimizes or neutralizes the electric field strength [2] in any direction that is not pointing to the center of the nucleon.

Since, a single neutron is unstable and has a half-life of about 14.9 minutes, one infers that a 2 down quark 'confinement' is unstable, as the proton's single down quark is stable and its 2 up quark confinement is stable, too. However, the 2 up quark *particle dielectric constant* $k_{c,Q}$=-1.939188x10^{-18} is greater than the 2 down quark *particle dielectric constant* $k_{c,Q}$=-4.912652x10^{-19}. This suggests that VEP matter is stable when

$$-1.958915x10^{-18} \le k_{c,0} \le 2.643848x10^{-19} \qquad (34)$$

and there is more to the quark particle structure (more elementary particles?) than is currently known.

With a Non *Standard Model (NSM)*, approach to nuclear forces the *Standard Model (SM)* is falsifiable. Unlike SM's "hideously complex . . . infinity", this approach only requires a finite amount of VEP matter to *confine* the nucleus.

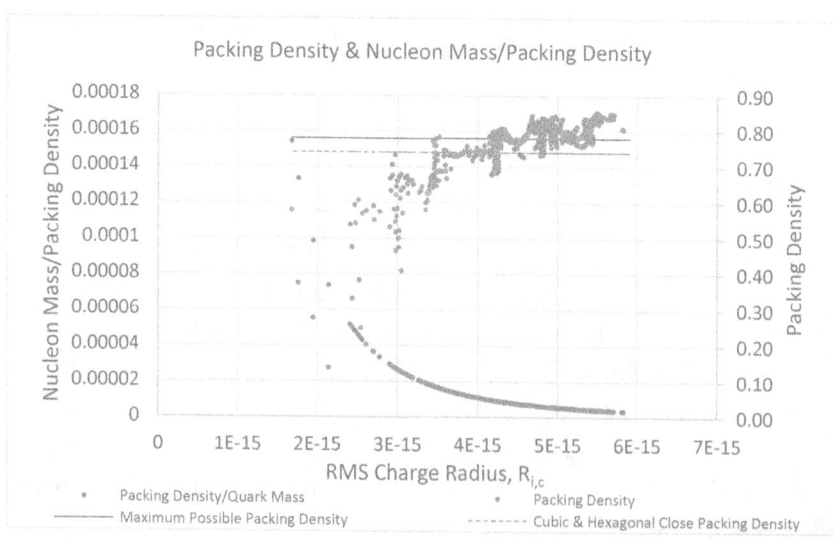

Fig. 3, Nuclei Packing Density Analysis

4.6 Do Nucleons Retain Structure?

Is it possible to determine some of the properties of the nuclei's structure from the empirical data alone? One possibility is to test if protons and neutrons retain their structure within a nuclei, as the Quark Velocity model suggests that they don't.

An approach to gauge this is to determine the nucleon packing density, η. If spheres are incompressible, then packing density η, the ratio of volume of spheres V_s to volume these spheres fill V_f, is given by $\eta = V_s/V_f$. Assuming that protons & neutrons are essentially spherical, the maximum theoretical packing density is $\eta_{max} \approx 0.77963557$, while the maximum with cubic and hexagonal close packing is $\eta_{ch} = 0.74048049$.

The spherical size of nuclei is determined from their RMC charge radius $R_{i,C}$. However, the neutron does not have an RMS charge radius, and an estimate $R_N = 2.857578 \times 10^{-45}m$ was determined by using the ratio of the proton-neutron Compton wavelengths, given a proton RMS charge radius $R_P = 2.845794 \times 10^{-45}$m. This neutron radius $R_N = 2.857578 \times 10^{-45}$m is only 0.41% larger than the proton's, and for packing density analysis can be assumed to be the same. A nuclei's empirical packing density $\eta_{i,E}$ was determined by equation (35), ratio of the sum of the proton-neutron spheres to the sphere of the nuclei,

$$\eta_{i,E} = \left(N_{i,P}R_P^3 + N_{i,N}R_N^3\right)/R_{i,C}^3 \qquad (35)$$

Where $N_{i,P}$ and $N_{i,N}$ are the number of protons and neutrons, respectively, in the nucleus of isotope i with RMS charge radius $R_{i,C}$.

Fig. 3 depicts (orange dots) the noisy packing density $\eta_{i,E}$ for 798 isotopes. It shows that isotopic empirical packing densities $\eta_{i,E}$ are greater than theoretical. 483 (60.5%) isotopes have $\eta_{i,E} > \eta_{max}$ or 0.77963557, and 663 (83.1%) have $\eta_{i,E} > \eta_{ch}$ or 0.74048049.

There are three possible explanations:

i) *Compression:* Protons and neutrons are compressible. This is not possible as the isotopic empirical packing densities $\eta_{i,E}$ are based on fixed nucleon sizes.

Unlike conventional packing density analytics, one is interested in (for want of a term) *tight packing density* of N number of inner spheres of volume V_s (proton & neutron spheres) that would *fit into the smallest outer spherical fill volume* V_f (nuclei size). Table 1(a) shows the *tight packing densities* of 2 spheres (that form a plane), 3 spheres (that form a plane) and 4 spheres (that form a tetrahedron). The fill volume V_f is a sphere whose radius reaches to the furthest outer edge of any of the

Table 1(a), Nucleon Theoretical Packing Densities

# of Spheres, of Radius $R_s=1$	Sphere Volume	Fill Volume Radius	Fill Volume	Theoretical Tight Packing Density $\eta_{i,T}$
	Vs	Rf	Vf	$\eta_{i,T}=Vs/Vf$
2	8.3776	2.0000	33.5103	0.2500
3	12.5664	2.1547	41.9034	0.2999
3	12.5664	2.1547	41.9034	0.2999
4	16.7552	2.2250	46.1401	0.3631

Table 1(b), Nucleon Theoretical & Empirical Packing Densities

# of Spheres	Isotope	Empirical RMS Charge Radius $R_{i,C}$	Empirical Packing Density $\eta_{i,E}$	Theoretical Tight Packing Density $\eta_{i,T}$
2	^{2}H	2.1402×10^{-15}	0.1389	0.2500
3	^{3}He	1.9448×10^{-15}	0.2775	0.2999
3	^{3}H	1.7591×10^{-15}	0.3755	0.2999
4	^{4}He	1.6757×10^{-15}	0.5787	0.3631

inner N spheres. Table 1 shows that the empirical packing densities $\eta_{i,E}$ for ^{2}H and ^{3}He are smaller than *tight theoretical* $\eta_{i,T}$. This suggests that these two nuclei occupy less volume than *tight theoretical*, while ^{3}H and ^{4}He occupy more. An inconsistency with so few protons and neutrons. Noting that Nature is consistent everywhere [2], this inconsistency is a result of a shrinking of nuclei size. For smaller nuclei the fill volume V_f is not the smallest theoretical spherical volume, i.e. tight packing is not so tight. This implies that the *VEP Cavity (VCav)* that contains the nucleus is determined by more than on factor.

ii) **Pass Through:** That electrically neutral neutrons, allow protons to pass through them without significant interaction. This facilitates packing into a smaller space and gives the appearance of compression. This hypothesis contradicts the data in Table 1, as ^{3}He with 1 neutron has smaller packing density than ^{3}H with 2 neutrons.

iii) **Cease to Exists:** A most likely scenario is that protons and neutrons cease to exist in their original form once a nucleus forms but exist as their constituent quarks. Thereby allowing for different arrangement within the nuclei. This would be consistent with the Quark Velocity model, and would explain why neutrons are apparently stable within a nucleus. It is because they cease to exist as a nucleon.

Thus, empirical packing densities suggests that nucleons, when in a nuclei, do not retain their structure i.e. nucleons only exist outside of nuclei. This would mean that the nucleus packing density is derived from quarks due to significantly different quark sizes i.e. smaller quarks can pack within the spaces of the bigger quarks and therefore result in the higher maximum empirical packing density $\eta_{i,E}$ of 0.85461598.

This begs the question what are the quark sizes? As a first pass to determining quark RMS charge radius, a) assume a particle's RMS charge radius is proportion to its charge; b) use the theoretical tight packing density for 3 particles (Table 1) of 0.2999 and increment this $\eta_{i,T}$ by 118.54% which is the adjustment required from $\eta_{i,T}$ to $\eta_{i,E}$ (see next section); c) using equation (35) for a proton gives, an Up Quark RMS charge radius $R_{i,U} = 3.242296 \times 10^{-16}$ and a Down Quark RMS charge radius $R_{i,D} = 1.621148 \times 10^{-16}$. These values are 36.88% and 18.44% that of the proton RMS charge radius, respectively.

4.7 The Role of Packing Density

The shrinking of the nuclei size implies other factors at work. Searching for an empirical packing density relationship across 765 isotopes, the blue dots in Fig. 3, gives an $R^2 > 99.9999\%$ equation (36).

$$\eta_{i,E} = k_{P,E} M_i / R_{i,C}^3 = k_{P,E} \rho_i \tag{36}$$

Where *empirical packing density constant* $k_{P,E} = 7.253246 \times 10^{-49}$, for isotope i of mass M_i (MeV). A new structural relationship exists, that there is a linear relationship between empirical packing density $\eta_{i,E}$ and mass density ρ_i. Further, the small model errors affirms that empirical packing density $\eta_{i,E}$, based incompressible quark sizes, is a very good model. Referring back to 6.i) therefore, packing density is a factor that determines $VCav$ *size*.

Since packing densities are about the number of particle spheres and not particle mass, using the data in Table 1(a) a regression model ($R^2 = 99.913\%$) was constructed for tight (theoretical) packing densities $\eta_{i,T}$. A maximum condition was imposed that at some large *Atomic Number* A_i say *1,000* the tight packing density should not exceed $\eta_{i,T} \leq \eta_{max}$ or 0.77963557. With this small sample size, an additional condition was included, that $\eta_{i,T} \leq \eta_{cb}$ or 0.74048049 at an Atomic Number A_i that would strengthen the regression relationship. By trial and error this was found to be $A_i = 500$. Thus, the tight (theoretical) packing density $\eta_{i,T}$ was found to be,

$$\eta_{i,T} = \sqrt{k_T ln(A_i)} = \sqrt{k_T ln(Z_i + N_i)} \tag{37}$$

Where the *tight packing density constant* k_T=0.089132.

Testing equation (37) against equation (35) showed it was sufficient to linearly translate (37) to (35) with a constant term k_{ssv} to minimize the differences between the two equations. This term k_{ssv} would translate *similar nucleon* sphere sizes (37) to *dissimilar quark* sphere sizes (35), giving,

$$\eta_{i,T} = k_{ssv}\sqrt{k_T \ln(A_i)} \tag{38}$$

Where the *sphere size variability constant* k_{ssv}=1.185444. Since both are physical properties, it should be possible to use *tight packing densities* $\eta_{i,T}$ to determine the nuclei's RMS charge radii $R_{i,C}$. Searching the data gives,

$$R_{i,C} = e^{k_P \eta_{i,T} + k_{PS}} \tag{39}$$

Where the *packing constant* k_P=2.317586 and *packing size constant* k_{PS}=-34.377521. Equation (39) has an R^2 of 99.408%. The negative *packing size constant* k_{PS}=-2.091362 is significant. A *possible interpretation* is as a sphere size variability term. The larger the variability or magnitude of $|k_{PS}|$ the smaller the radius $R_{i,C}$

Searching for an equation similar to (36) gives,

$$\eta_{i,T} = k_{P,T}M_i/R_{i,C}^3 = k_{P,T}\rho_i \tag{40}$$

Where *tight packing density constant* $k_{P,T}$=8.746260x10^{-49}, for isotope *i* of

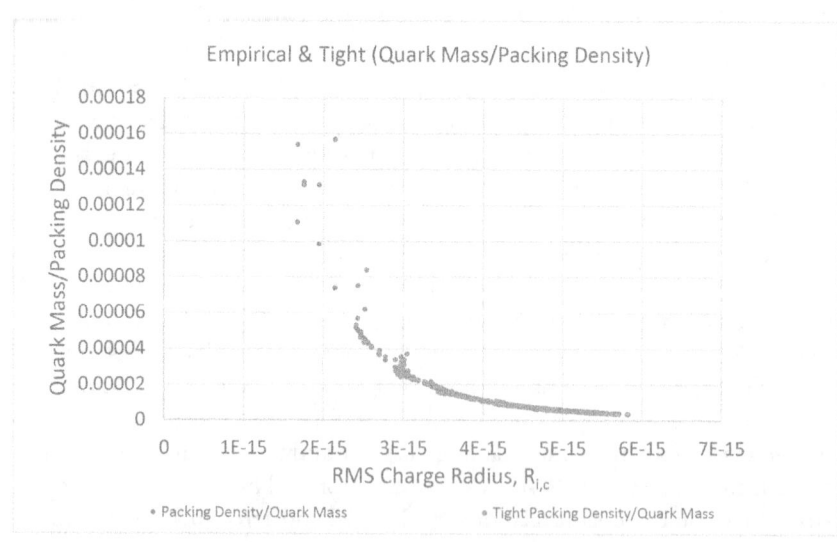

Fig. 4, Differences between Empirical & Tight Packing Densities for 765 isotopes

mass M_i (MeV). Fig. 4 shows this is not satisfactorily good relationship as $R^2=88.856\%$. A significantly better $(R^2=99.437\%)$ relationship is

$$\left(\eta_{i,E}/M_i\right) = k_{PC}\left(\eta_{i,T}/M_i\right)^{k_{PE}} \tag{41}$$

Where *packing coefficient* k_{PC} = 1.694662 and *packing exponent* k_{PE} = 0.990871. Since $k_{PE} < 1$, the ratio of empirical packing density $\eta_{i,E}$ to isotopic mass M_i, does not increase as quickly as the ratio of tight packing density $\eta_{i,T}$ to isotopic mass M_i. As isotopic mass is also a measure of number of nucleons, up & down quarks and therefore electric charges, this suggests that electrical forces slow the rate at which empirical packing densities increase. Using particle count N_i as a proxy for amount electric charge, instead of mass, in the denominator of this ratio, the up quark $(R^2=99.405\%)$ gives a *packing exponent constant* k_{PE} = 0.991817 closest to 1 and *packing coefficient* k_{PC} = 0.967067,

$$\left(\eta_{i,E}/N_i\right) = k_{PC}\left(\eta_{i,T}/N_i\right)^{k_{PE}} \tag{42}$$

Thus, one concludes that the new structural relationship, packing density is a significant determinant of nucleus mass, size and mass density.

To make sure nothing was missed, this isotopic data was structured to include only 615 (of the 758) neighboring isotopes i.e. *Atomic Number increase* δA_i by *1* or $\delta A_i=1$. What stood out as significant across all neighboring isotopes was the ratio of this change in the neighboring RMS charge radius $\delta R_{i,c}=R_{i,C} - R_{i-1,C}$ to the RMS charge radius $R_{i,C}$. Fig. 5 and Table 2, shows 4 distinctive straight lines. At this time one can only guess that these relationships are symptomatic of an unknown underlying nuclei packing structure.

4.8 Role of Binding Energy

Binding energy $E_{i,B}$ (MeV) of isotope i is expressed as the difference between sum of masses of the individual nucleon and the combined mass of the nucleus. Therefore, it should be related to packing density. Searching the data gives a binding energy $E_{i,B}$ relationship (R^2 of 99.655%),

Series	Min A|Z|N	Max A|Z|N	Count	R^2	Slope
A	7|3|4	47|20|27	37	99.29%	3.716E+14
B	57|26|31	139|62|77	185	99.49%	3.054E+14
C	109|49|60	246|91|155	353	99.76%	2.261E+14
D	25|12|13	55|25|30	40	99.59%	1.862E+14

Table 2: Ratio of $\delta R_{i,c}=R_{i,C} - R_{i-1,C}$ to $R_{i-1,C}$

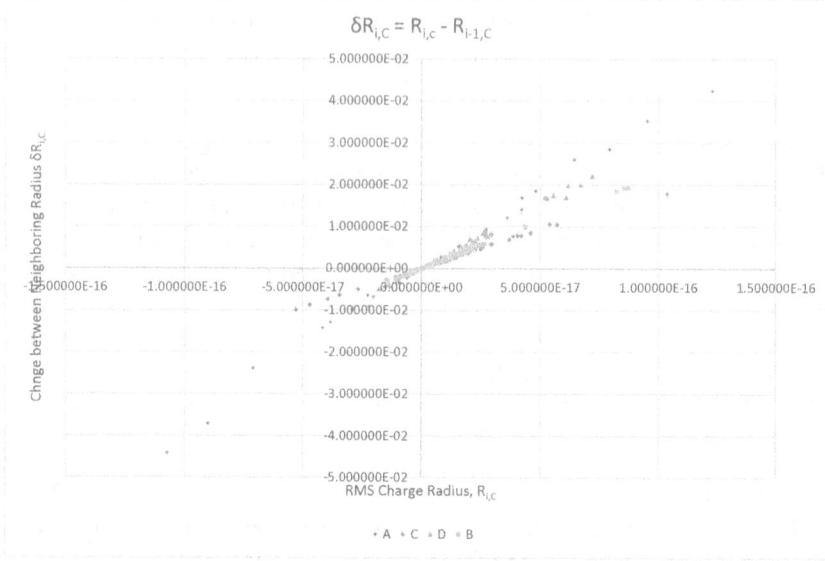

$$\delta R_{i,C} = R_{i,c} - R_{i-1,C}$$

Fig. 5, $\delta\, R_{i,c} = R_{i,C} - R_{i-1,C}$ versus $R_{i,C}$

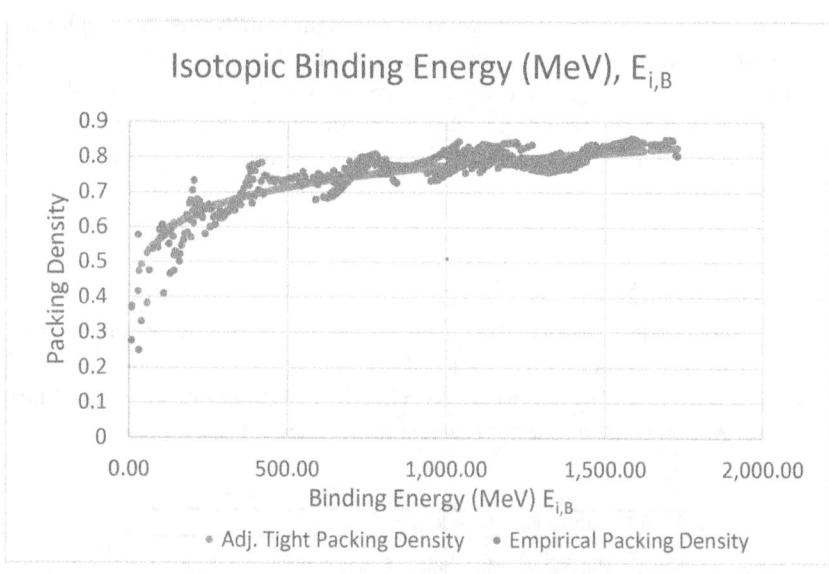

Fig. 6, Isotopic Binding Energy as a function of Tight Packing Densities

$$E_{i,B} = e^{k_B \eta_{i,T} + k_{ER}} \tag{43}$$

Where the *binding constant* k_B=*11.593234* and *energy ratio constant* k_{ER}=-*2.091362*.

Fig. 6 shows that *tight packing density* $\eta_{i,T}$ is a significantly better predictor of binding energy than *empirical packing density* $\eta_{i,E}$.

The negative *energy ratio constant* k_{ER}=-*2.091362* is significant. A *possible interpretation* is as a sphere size variability term. This term decreases the amount of binding energy required as the sphere size variability increases,

$$E_{i,B} = e^{k_B \eta_{i,T} + k_{ER}' k_{ssv}} \tag{44}$$

Where the *adjusted energy ratio constant* k'_{ER}=-*0.566828*.

The new structural inference is that the structural integrity of the nucleus is derived from tight packing densities, and binding energy $E_{i,B}$ is consumed to achieve the necessary *tight packing densities* $\eta_{i,T}$.

4.9 VEP Cavity (VCav)

Let's recap. Nucleons cease to exist within a nuclei. The nucleus, like nucleons, consists of quarks of specific spherical sizes. These sizes determine packing densities. Binding energy $E_{i,B}$ (43) is utilized to maintain *tight packing density* $\eta_{i,T}$, (37). This in turn determines the *empirical packing density* $\eta_{i,E}$ (43) and therefore the mass densities ϱ_i (36) of nuclei.

Since Poincaré stresses [13] do not exist [2] electric field lines are not repulsive. VEP matter maintains mechanical and electrical stability by converting electric forces to centripetal forces. As quarks carry all the mass of nucleons and nuclei, VEP matter is massless. It is therefore proposed that VEP matter is a field similar to spacetime. With very different dielectric properties to that of spacetime, VEP matter therefore, replaces spacetime in nucleons and nuclei. This difference between VEP matter and spacetime explains velocity refraction (8) at the RMS charge radial boundary $R_{i,C}$. The resulting velocity $v_{s,t}$ due to this diffraction is the starting edge of gravitational fields [1].

The dielectric constant κ of VEP matter decreases with the RMS charge radius (22), (23), (31) & (32). Assuming that these 'orbits' are randomly oriented, then the isotope's RMS charge radius is determined by the *tight packing density* $\eta_{i,T}$, equation (39) and with equation (15),

$$\kappa_Q = {}^{(-k_P/k_B)} \sqrt{E_{i,B}} \left(k_{c,Q} e^{k_{PS} + k_{ER} - k_P/k_B} \right) \tag{45}$$

79

or

$$K = k_{c,0}e^{k_{PS}+k_{ER}k_P/k_B} = k_{c,0}7.73503x10^{16} \tag{46}$$

and,

$$\kappa_Q = K^{(-k_P/k_B)}\sqrt{E_{i,B}} = K^{(-0.2)}\sqrt{E_{i,B}} \tag{47}$$

That is, VEP matter is from the binding energy that was originally derived from mass. VEP matter responds to electric forces in the nuclei to instantiate centripetal forces. These centripetal forces are seen as packing and mass densities. The *VEP Cavity (VCav)* is formed by the packing of quarks in VEP matter.

4.10 Conclusion

By eliminating the need for weak and strong nuclear forces, this paper has shown that the *Standard Model (SM)* is falsifiable. Two key concepts, supported by the empirical data of 798 isotopes, facilitated this falsifiability. These two concepts are *Variable Electrical Permittivity* (VEP) matter and packing densities.

References:

[1] Solomon, B. T., "The Variable Isotopic Gravitational Constant", in *Super Physics for Super Technologies*, March 2015..

[2] Solomon, B.T., "A Universal Approach to Forces", in *Super Physics for Super Technologies*, March 2015..

[3] Solomon, B. T., "Empirical Evidence Suggest A Need For A Different Gravitational Theory," in the proceedings of the *100 Year Starship Study Public Symposium (100YSS,)*, 2013.

[4] Solomon B.T., "New Evidence, Conditions, Instruments & Experiments for Gravitational Theories", Journal of Modern Physics, Special Issue on Gravitation, Astrophysics and Cosmology, Vol. 8A, 2013, August 2013.

[5] Solomon, B. T., "Empirical Evidence Suggest A Need For A Different Gravitational Theory," *American Physical Society (APS) April Conference, Denver*, 2013

[6] Solomon B.T., *An Introduction to Gravity Modification: A guide to using Laithwaite's and Podkletnov's experiments and the physics of forces for empirical results.* Universal Publishers, Boca Raton, 2nd Edition, May 2012.

[7] Solomon, B. T., "Non-Gaussian Radiation Shielding," *100 Year Starship Study Public Symposium (100YSS,)*, 2011.

[8] Solomon, B.T., "Gravitational Acceleration Without Mass And Noninertia Fields", Physics Essays, Vol. 24, 327, 2011. [Phys. Essays **24**, 327 (2011)]

[9] Solomon, B. T., "Reverse Engineering Podkletnov's Experiments," in the proceedings of the *Space, Propulsion & Energy Sciences International Forum (SPESIF-11)*, Edited by Glen A Robertson, Physics Procedia, Elsevier Science.

[10] Solomon, B. T., "Non-Gaussian Photon Probability Distributions," in the proceedings of the *Space, Propulsion & Energy Sciences International Forum (SPESIF-10)*, Edited by Glen A Robertson, AIP Conference Proceedings **1208**, Melville, New York, (2010).

[11] Solomon, B. T., "An Approach to Gravity Modification as a Propulsion Technology," in the proceedings of the *Space, Propulsion & Energy Sciences International Forum (SPESIF-09)*, Edited by Glen A Robertson, AIP Conference Proceedings **1103**, Melville, New York, (2009).

[12] Solomon, B. T., "An Epiphany On Gravity", Journal of Theoretics, Vol. 3-6, 2001.

[13] Feynman, R., *Feynman Lectures on Physics*, Chapter 28-4.

[14] Nemiroff, R., "Bounds on Spectral Dispersion from Fermi-Detected Gamma Ray Bursts," Physical Review Letters, Vol. 108, No. 23, 2012.

[15] Cho, A., Mass of the Common Quark Finally Nailed Down, Science Magazine, 2 April, 2010.

[16] Davies, 1.C. T. H., McNeile, C. , Wong, K. Y., Follana, E., Horgan, R., Hornbostel, K., Lepage, G. P., Shigemitsu, J., Trottier. H., Precise Charm to Strange Mass Ratio and Light Quark Masses from Full Lattice QCD. Physical Review Letters, 2010; 104 (13): 132003 DOI: 10.1103/PhysRevLett.104.132003

[17] Angeli, I., "A consistent set of nuclear rms charge radii: properties of the radius surface (R,N)", Science Direct, Volume 87, Issue 2, July

Super Physics for Super Technologies

2004, Pages 185–206, DOI: 10.1016/j.adt.2004.04.002

End Notes:

i.   Does not include non-peer reviewed conference presentations between 2001 and 2009.

ii. Macro body elongation due to tidal gravity is attributed to Roger Penrose (this author could not find the reference in time for this paper). Macro bodies elongate as the body falls into a gravitational field. Let's reexamine this tidal behavior with the additional requirement that this tidal gravity property be consistent with Lorentz-FitzGerald transformations or Special Theory of Relativity. To be consistent with Lorentz-FitzGerald transformations, the atoms and elementary particles would contract in the direction of the fall. However, to be consistent with tidal gravity's elongation, the distances between atoms in the macro body has to increase at a rate consistent with the acceleration and velocities experienced by the various parts of the macro body. That is, as the atoms get flatter, the distances apart get longer. One suspects that this axiom's inconsistency with the empirical evidence has led to an explosion of string theories, each trying to explain Nature with no joy.

iii. Accepting that a single neutron will decay.

iv. Standard Model proposal.

v.   http://hyperphysics.phy-astr.gsu.edu/hbase/particles/proton.html

vi. http://mathworld.wolfram.com/SpherePacking.html

vii. http://physics.nist.gov/cuu/Constants/index.html

viii. Based on 765 isotopes that match [17] with documented AMUs per spreadsheet prepared by Dr. Gordon Gilmore of Nuclear Training Services Ltd. Data taken from the 1995 update to the atomic mass evaluation by G.Audi and A.H.Wapstra, Nuclear Physics A595 vol. 4 p.409-480, December 25, 1995. http://physics.nist.gov/PhysRefData/Compositions/mass_rmd.mas95round.txt

ix. The Pareto of Errors are 0 points $>|\pm5.0\%|$, 0 points $>|\pm2.5\%|$, 0 points $>|\pm1.0\%|$, 0 points $>|\pm.5\%|$ & 1 point $>|\pm0.1\%|$

x.   The Pareto of Errors are 6 points $>|\pm5.0\%|$, 25 points $>|\pm2.5\%|$ & 326 points $>|\pm1.0\%|$

xi. Exclude 8 light isotopes of H, He, Li, Be & Ne.

xii. The Pareto of Errors are 129 points $>|\pm5.0\%|$, 394 points $>|\pm2.5\%|$ & 656 points $>|\pm1.0\%|$

xiii. The Pareto of Errors are 11 points $>|\pm5.0\%|$, 69 points $>|\pm2.5\%|$ & 425 points $>|\pm1.0\%|$

xiv. The Pareto of Errors are 16 points $>|\pm5.0\%|$, 91 points $>|\pm2.5\%|$ & 467 points $>|\pm1.0\%|$

5. Replacing Schrödinger

Abstract:- This paper proposes the *Probabilistic Wave Function* Ψ_P that is different from Schrödinger's Ψ_S by reversing engineering the empirical data and changing the underlying physical process.

It is found that though both solutions look similar, the Schrödinger wave packet is a damped oscillation while the *Probabilistic* is an inverse function of distance.

Its solution unlike Schrödinger's is non integrable thereby suggesting that conservation of mass-energy is not a necessary requirement to arrive at a solution.

Probabilistic Wave Function solution ψ_P consists of two components, *space wave* χ_P and *envelope function* φ_P. How these components interact with their environment explains interference, entanglement and subwavelength confinement.

Several new double slit experiments are proposed to test the validity of these proposals. Finally, a first attempt at the new *Orbital Space Wave* electron shell is proposed. It has excellent agreement with the quantum mechanical electron shell, and needs further development.

5.1 Premise

This paper reports the 14th paper in the 16-year ([1] to [13][i]) investigation into the feasibility of gravity modification, and a *New Standard Model*, proposing:

i)   Non Inertia (Ni) Fields defined as the *spatial* gradient of real or virtual velocities.

ii) Equation (1), the massless formula, for gravitational acceleration, where τ is the change in time dilation divided by the change in that distance,

$$g = \tau c2 \qquad (1)$$

iii) All macro forces (electromagnetic, electric & magnetic, mechanical, gravitational), can be model by the spatial gradient of their fields [3].

iv) The isotopic gravitational constants that explains how matter, and not mass, causes gravitational fields.

v)   The falsifiability of the Standard Model.

vi) And an approach to replacing the Schrödinger wave function (this paper).

As a consequence of iii), that electric and magnetic field lines are not repulsive, Poincaré stresses [14] do not exist [3] thus facilitating simpler proton, neutron & atomic nuclei models [1].

Occam's Razor, that the simpler solution is most likely the correct solution, drives this investigation for a simpler matter model that explains how matter creates gravity.

5.2   Introduction

Starting with Nemiroff's [14] finding, that quantum foam cannot exist in free space, evidence is mounting that theoretical physics is in need of a substantial rewrite, may be even a complete overhaul. Solomon proposed [3] three important empirical based requirements of theoretical physics. (See [5] for the rest of his proposals.) These are:

i)   ***Non-Gaussian Probability:*** The [4], [5], [6], [7], [8] & [11], Airy disc[ii] from an aperture diameter of w_A, has an intensity pattern I with reference intensity I_0, at a point on an orthogonal radial distance r_s, on a screen at a distance d_s. This empirical evidence (2) shows that photon probability is not Gaussian.

$$I=(I_0 \sin(u))/u \qquad (2)$$
$$\theta=atan(r_s / d_s) \qquad (3)$$
$$u=\pi(w_A / \lambda)\sin(\theta) \qquad (4)$$

This photon probability is the new modified Gamma distribution, named *Var-Gamma* (5) because this distribution's a & β parameters are variable functions of orthogonal distance r_s from the axis of photon propagation of wavelength λ.

$$(r) = \frac{1}{\beta\Gamma(\alpha)}\left(\frac{r}{\beta}\right)^{\alpha-1} e^{-r/\beta}$$

(5)

$$\alpha = r_s$$ (6)

$$\beta = r_s / \sqrt{u}$$ (7)

The mass particles' (e.g. electrons) Airy discs are consistent with massless photon Airy discs. Therefore, this *Var-Gamma* probability function holds for mass-based particles, and there is a consistency in particle structure design between mass & massless particles.

ii) **Compressive Particles:** Equation (1) was discovered [4], [5], [6], [7], [8], [9] & [12] on the basis that particles are compressive and shorten in the direction of propagation/travel. A consistency with Lorentz FitzGerald transformations (LFT). This is unlike quantum or string theories [2] which require particles to be point like or expand with energy[iii], respectively. Therefore, the new finding $g=\tau c^2$ affirms that particles are compressive.

iii) **Inconsistent Wave Function:** All particles, with and without mass, have wave functions that spread out into the region of space surrounding the particle. In single and double slit experiments, particles exhibit a consistent wave interference pattern, irrespective of whether they have mass or not. This suggests, that both photons and mass-based particles have identical wave function mechanisms irrespective of mass.

However, photons velocity v_p is the velocity of light $(v_p=c)$ while mass-based particle velocity v_p is less than that $(v_p<c)$. To be consistent with Lorentz-FitzGerald transformations (LFT) and Special Theory of Relativity (STR), anything traveling at the velocity of light must have zero thickness, in the direction of propagation, and cannot spread out like the wave function does.

Therefore, the logical resolution is that the wave function is not moving in the direction of propagation or travel. Its velocity $v_{wf} = 0$, is zero, and therefore, independent of v_p.

$$v_{wf} \neq f(v_p)$$ (8)

A zero velocity $v_{wf} = 0$ wave function is consistent with both types of particle velocities $v_p<c$ and $v_p=c$.

How could Nature implement such a property? Here is an analogy. Take a garden rake, turn it upside down and place it under a carpet. Move it. What does one observe?

The carpet exhibits a wave-function-like-envelope-bulge that appears to be moving in the direction the garden rake is moving.

But the bulge is not moving. It shows up wherever the garden rake is. The rake is moving but not the bulge. The bulge is simply an orthogonal displacement of the carpet caused by the rake.

The wave function, like the carpet bulge, is an orthogonal displacement in spacetime caused by the presence of the particle, and is not moving. The wave function spreads across the spacetime in a consistent manner whether it is a photon or a mass-based particle, and therefore, the wave-particle duality no longer exists in the manner expressed in contemporary physics.

This zero-velocity bulge-like wave function is now consistent with Einstein's Special Theory of Relativity (STR) and with the empirical evidence of the Lorentz-FitzGerald transformations (LFT).

The *Standard Model (SM)* is successful because, just as the shape of the carpet bulge is unique to the shape of the garden rake, so is the wave function displacement in spacetime unique to the properties of the underlying particles.

To resolve i) & iii) by reversing engineering the empirical data, this paper proposes a wave function Ψ that is different from Schrödinger's[iv], Ψ_S (9) with their respectively solutions ψ and ψ_S and quantum mechanical spatial solution ψ_{SS}

$$\psi_S(x,t) = \psi_{SS}(x)[cos(\omega t) - isin(\omega t)] \tag{9}$$

$$\psi_S(x,t) = e^{-i\omega t}\psi_{SS}(x) = e^{-i\omega t}Asin(k_\psi x) \tag{10}$$

$$\psi_{SS}(x) = Asin(kx) \tag{11}$$

$$k_\psi = 2\pi/\lambda \tag{12}$$

Where A is the maximum amplitude, $Asin(k_\psi x)$ is the amplitude between nodes at distance x, and ω is angular frequency of this standing wave.

5.3 Non-Gaussian Probability

The Var-Gamma probability distribution (blue line, Fig. 1) proposed in [4], [5], [6], [7], [8] & [11] is based on the assumption that the absolute magnitude of the wave function oscillates (red dashed lines, Fig. 1) about this Var-Gamma probability function.

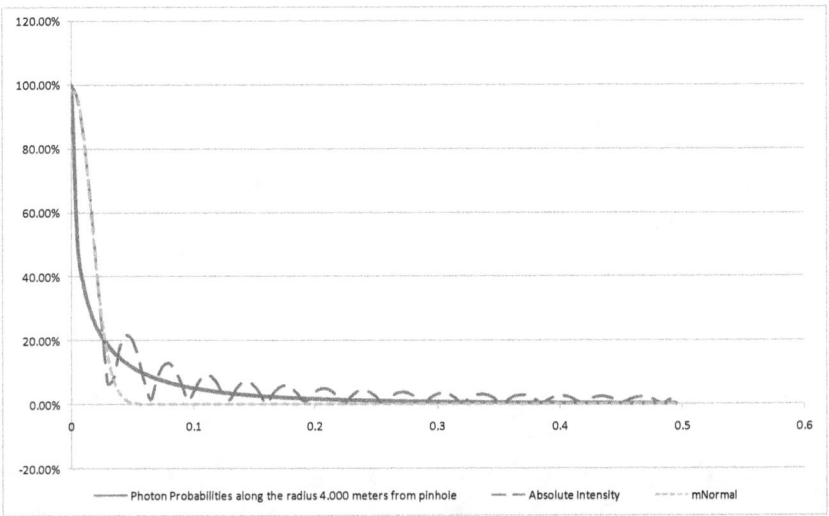

Fig. 1, Construction of the Var-Gamma probability distribution from the Airy pattern

A second approach is now proposed. Fig. 2 (red line) shows the *solution* ψ (13) to the *Probabilistic Wave Function* Ψ,

$$\psi = I/I_0 \tag{13}$$

This *probabilistic wave solution* ψ is constructed from 2 components. The first, is the wave term in spacetime, the *space wave* χ (Fig. 2, dotted grey line), (14)

$$\chi = sin(u) = sin\big(\pi(w_A/\lambda)sin(\theta)\big) \tag{14}$$

And the second is the *envelope function* φ (Fig. 2, dashed green line) of *unnormalized* probabilities from the u function of (4), as follows,

$$\varphi = 1/u = 1/\big(\pi(w_A/\lambda)sin(\theta)\big) \tag{15}$$

Such that,

$$\psi = I/I_0 = \chi\phi = sin(u)/u \tag{16}$$

In effect, the wave *space wave* χ (14), weights the probability *envelope function* φ (15), to produce the *probabilistic wave solution* ψ (16). This weighting is equivalent to the *space wave* χ casting a *shadow* on the *envelope function* φ and upper bounds the *probabilistic wave solution* ψ in a consistent mathematical manner.

Both approaches (5) and (15) to determining the *empirical* particle proba-

87

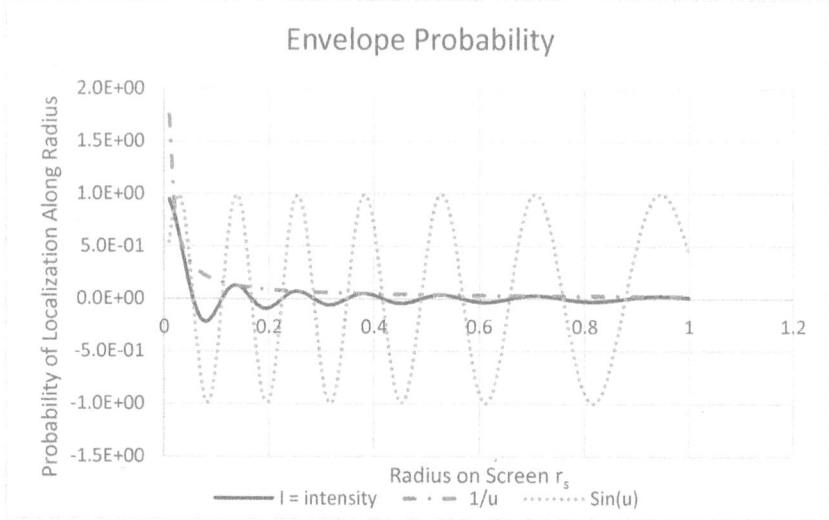

Fig. 2, Probabilistic wave solution ψ (red) & its 2 components, χ (grey) & φ (green)

bility distribution result in non-Gaussian probability functions; and (15) is simpler and easier to use than (5).

5.4 Wave Function Deconstruction

Space wave χ (Fig. 2, dotted grey line) has a constant amplitude (maximum of $I/I_0=1$) and damped period. This is because the *space wave* χ wavelength $λ_χ$ increases as the radial distance r_s increases.

Compare (10) with damped oscillations[v] (17).

$$x = e^{-γt} a\cos(ω_1 t - α) \tag{17}$$

Where $γ$ is the damping coefficient, a amplitude, $α$ is phase, and $ω_1$ is related to damped oscillations.

Clearly, the Schrödinger's wave function solution $Ψ_S$ (10) is a damped oscillation that is phase shifted.

Replacing the inverse function *envelope function* φ with an equivalent exponential[vi] term (18), and the time t variable with radial distance r_s gives (19), a damped oscillation version of ψ.

$$φ = e^{-γr_s} \tag{18}$$

$$ψ = e^{-γr_s} a\cos(ω_1 r_s - α) \tag{19}$$

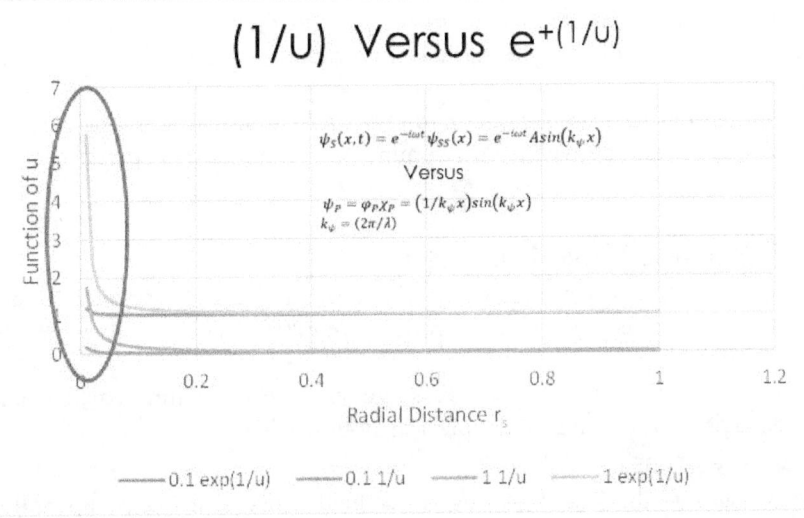

$$\psi_S(x,t) = e^{-i\omega t}\psi_{SS}(x) = e^{-i\omega t}A\sin(k_\psi x)$$

Versus

$$\psi_P = \varphi_P \chi_P = (1/k_\psi x)\sin(k_\psi x)$$
$$k_\psi = (2\pi/\lambda)$$

Radial Distance r_s

——— 0.1 exp(1/u) ——— 0.1 1/u ——— 1 1/u ——— 1 exp(1/u)

Fig. 3, Test for nature of the Envelope Probability Function for $d_s=0.1m$ and $1.0m$

Fig. 3 shows that both the inverse (15) and the exponential (18) forms of the *envelope function* φ term appear similar. Even though they look similar, the exponential term (18) results in a much higher wave packet mode than the inverse term. The exponential term (18) is at best a good approximation[vii] but not a perfect fit. Therefore, neither the *space wave* χ nor the *probabilistic wave solution* ψ are damped spatial oscillations.

However, both the *space wave* χ and the *probabilistic wave solution* ψ have damped *spatial* periods, as determined by (14).

Equation (14) shows that the *space wave* χ, is constant when θ is constant, that is, where the ratio of r_s to d_s is constant,

$$k_\theta = sin(atan(r_s/d_s)) \tag{20}$$

$$\chi = sin(k_\theta(w_A/\lambda)\pi) \tag{21}$$

As a function of θ, (20) confirms that the *space wave* χ, is not a time varying standing wave function, $k_\theta(w_A/\lambda)\pi$ represents the phase shift of these spatial oscillations, and spacetime disturbance χ is the outward projection of the particle's own wave function.

By (15) & (20), (22) confirms that the *envelope function* φ is a spatial geometric structure whose probability decreases with the particle's wavelength λ.

$$\varphi = 1/(k_\theta(w_A/\lambda)\pi) \tag{22}$$

That is, the higher the energy of the particle, the smaller the *space wave χ*, and lower the probability of particle localization within this smaller geometry.

Therefore, the *space wave χ*, the *envelope function φ* and thus, the *probabilistic wave solution ψ* are derived from a common mechanism for all particles.

5.5 Particle Wave Function

Therefore, it is proposed that the shape of the particle's *probabilistic wave solution ψ*P can be reconstructed from its *space wave χ*P and its *envelope function φ*P.

Assuming the particle occupies some space, extrapolating back (along the z-axis) to the center of the particle, $d_s=0$ would give,

$$k_\theta = sin(90°) = 1 \qquad (23)$$

Along both the x- and y-axis, given x the distance from the center of the particle, the aperture $w_A=2x$. Given k_Ψ (11), by (21) particle's *space wave χ*P is,

$$\chi_P = sin(x2\pi/\lambda) = sin(k_\psi x) \qquad (24)$$

From (22) the particle's *envelope function φ*P is,

$$\varphi_P = \lambda/2\pi x = 1/k_\psi x \qquad (25)$$

And from (15) the particle's *probabilistic wave solution ψ*P is,

$$\psi_P = \varphi_P \chi_P = (1/[k_\psi x])sin(k_\psi x) \qquad (26)$$

The particle's *probabilistic wave solution ψ*P (26) explains how Airy and interference patterns are produced. The particle's *envelope function φ*P is the maximum probability amplitude and its *space wave χ*P determines the oscillations along the x- and y-axes. Note also, that the particle's *space wave χ*P (24) is not period damped like that of the interference patterns *space wave χ* (21). This affirms that the latter (21) is a projection of the former (24).

Since (26) is non integrable, it will not be easily to reverse engineer a Schrödinger-type wave function, if one exists. A non integrable *probabilistic wave solution ψ* (26) would concur with the thesis that this wave function is an orthogonal displacement in spacetime which is not motion.

Therefore, along the z-axis by (20), as θ→0

$$k_\theta = sin(atan(0/d_s)) \to 0 \qquad (27)$$

And as

$$\chi_{P,z} = sin(k_\theta(w_A/\lambda)\pi) \to 0 \qquad (28)$$

$$\varphi_{P,z} = 1/(k_\theta(w_A/\lambda)\pi) \to 1/0 \qquad (29)$$

However, $\theta \to 0$

$$\psi_{P,z} = \chi_p \phi_P \to k_\theta (w_A/\lambda)\pi / k_\theta (w_A/\lambda)\pi = 1 \qquad (30)$$

Equation (26) describes the particle's wave function $\Psi_P = \psi_P$ in both the x- and y-axes while the z-axis function (30) is a constant 1. It shows that, the particle wave function along the z-axis is not a function of this distance, and thus the wave function is orthogonal to the particle's motion. The particle's intensity shape is symmetrical conic-like disc in the plane orthogonal to motion, but its physical shape is an infinitely thin disc. This is a structure that would be consistent at any velocity $v_p < c$ and $v_p = c$.

The basic particle can be described as consisting of a wave function disc and probability disc that are orthogonal to its vector motion. Therefore, there are 5 parts to a basic particle, i) vector of motion, ii) *space wave* χ_P disc (24), iii) probabilistic *envelope function* φ_P disc (25) that iv) combine to form the *probabilistic wave solution* ψ_P disc (26) and v) the projected *probabilistic wave solution* ψ disc (16) that results in the diffraction patterns.

One last point. Looking at how the de Broglie wavelength was constructed, from the perspective of *space waves*, one can restate that the total *mass equivalent energy* $E_{m,T}$ of a particle with mass m traveling at a velocity v, in terms of its rest mass $E_{m,0}$ (the rest mass term) and its *kinetic energy mass* $E_{m,K}$ (the energy term) as,

$$E_{m,T} = E_{m,0} + E_{m,K} \qquad (31)$$
$$E_{m,T} = mc^2 + mv^2 \qquad (32)$$

And therefore, from (32), the total mass wavelength λ_T, the rest mass wavelength λ_0 and the kinetic energy wavelength λ_K are given by,

$$\frac{h}{\lambda_T c} = \frac{h}{\lambda_0 c} + \frac{h}{\lambda_K v} \qquad (33)$$

And

$$\lambda_0 = \frac{h}{mc} \qquad (34)$$

$$\lambda_K = \frac{h}{mv} \qquad (35)$$

That is, the energy term $h/\lambda_K v$ in (33) determines the *projected* wave function wavelength λ_K per (14) & (16) and the mass term $h/\lambda_0 c$ in (33) is the *true* wave function's wavelength λ_0 per (24) & (26) of the mass particle. However, since photons do not have mass, the *true* wave function's wavelength is also the *projected* wave function wavelength. Therefore, one notes that there two independent mass terms in the particle structure.

Though they appear similar the Schrödinger wave solution ψ_S and the particle's *probabilistic wave solution* ψ_P, have several differences,

i)   Unlike the Schrödinger wave solution ψ_S (10) where the probability is the square of the wave amplitude spatial ψ_S (11), *probabilistic wave solution* ψ_P has an amplitude φ_P, the *un-normalized* probability.

ii)   The amplitude of the Schrödinger wave solution ψ_S is a damped oscillation because it is an exponential function (10), unlike the amplitude φ_P (25) of the *probabilistic wave solution* ψ which is an inverse function.

iii) The *space wave* χ_P, the sine term of (24) has a constant period for a given λ and a geometric structure in distance x. Schrödinger's wave function solution ψ_S, however, is a time t based standing wave (9) & (10) or net stable effects of time based oscillations.

iv)   The Schrödinger wave packet comprises of multiple frequenciesv[iii] shaped as a damped oscillation. While the *probabilistic wave function's packet* ψ_P is of a single frequency with wavelength (period) λ and its wave packet shape is an inverse of distance x.

In effect, the Schrödinger wave function Ψ_S solution ψ_S can be written in *probabilistic wave solution* ψ form (25)

$$\psi_S(x,t) = \{A[cos(\omega t) - isin(\omega t)]\}sin(kx) \qquad (36)$$

That is, the two approaches have the same spatial function *sin(kx)* but differ otherwise.

Per 2iii), the proposed common mechanism is the particle's wavelength λ_P disturbs spacetime to generate orthogonal *spatial oscillations* λ_χ such that all three particle phenomena, *probabilistic wave solution* ψ_P, the *space wave* χ_P, and *envelope function* φ_P, are not moving and consistent with both STR and LFT. From (26),

$$\lambda_\chi = \lambda_P \qquad (37)$$

As a test (26) was plotted against the atomic mass (MeV) of 766 nuclei isotopes [18], using the de Broglie rest mass wavelength, and the RMS charge radius $R_{i,C}$ for x. See Fig. 4.

Fig 4. shows that the points (mustard colored dots) are spread all over the place due to the RMS charge radius $R_{i,C}$ not being multiples of 2π. However, they are bounded by the maximum *probabilistic wave solution* (green line). The analysis of this empirical data shows that the upper bound of (26) is given by (33),

$$\psi_{P,Upper\ Bound} = k_{UB}\lambda \qquad (38)$$

where *wave solution constant* $k_{UB} = 7.417188x10^{13}$. That is, the empirical data supports a probabilistic wave solution of the form of (26). Except for the proton, the empirical analysis also shows that the RMS charge radius $R_{i,C}$ is always greater than the de Broglie rest mass wavelength.

5.6 Effect of Spatial Probability

Given the deconstruction of the particle's *probabilistic wave solution* ψ_P into its two components the *space wave* χ_P and the *envelope function* φ_P, one can now test for φ_P in other experimental observations. Two important tests [8] are quantum entanglement and electron shell.

A note. For ease of use, this *envelope function* φ_P over the disc or volume of the particle's *space wave* χ_P is termed the *spatial probability distribution*, P_S. The total P_S must sum to *1*, at or in the *field of interaction* F_I irrespective of this field's shape. This *field of interaction* F_I is determined on a case by case basis. It is the area A of the plane as in the screen for Airy disc experiments (34) and the volume V of space (35) for 3-dimensional chemical reactions.

$$F_{I,A} = \int_0^L \pi\varphi_P^2 dr_s$$

(39)

Where r_s is from 0 or from the axis of propagation through the aperture, to the radial distance L on the screen. Ideally $L=\infty$ but in practice $L\approx15m$.

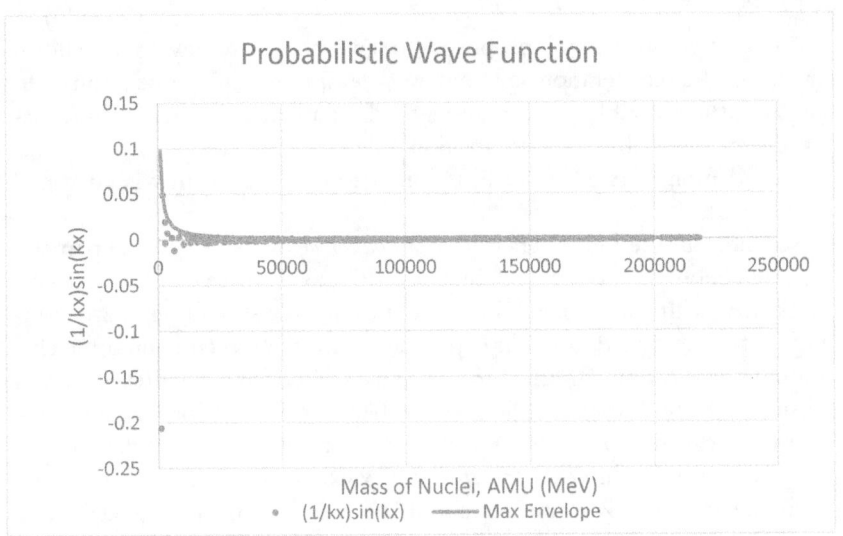

Fig. 4, The Probabilistic Wave Function for 766 isotopes.

$$F_{I,V} = \int_0^{d_s} F_{I,A} dz$$

(40)

Where the volume is formed by the sum of the discs $F_{I,A}$ along the z-axis, from the aperture at 0 to a distance d_s from this aperture.

To normalize an *envelope probability* $\varphi_{Pi,(x,y,z)}$ of photon i localizing at coordinates (x, y, z) on a screen a distance z, the *normalized probability* $P_{Ni,(x,y,z)}$ would take the form,

$$P_{Ni,(x,y,z)} = \varphi_{Pi,(x,y,z)}/F_{I,A}$$

(41)

The *space wave* χ_P does explain how a particle can interfere with itself. Since the *space wave* χ_P is a disturbance in spacetime, as this disturbance passes through both slits, it regenerates as two new *space waves* χ_{PL} and χ_{PR} at the left and right slits, and total normalized probabilities P_{NL} and P_{NR} from each slit, respectively. Having crossed the slits, the total probability of localization $P_{NL} + P_{NR}$ on the screen ahead must equal one. Therefore, the same interference pattern is observed as if two particles are present.

One could add that probabilities can be transformed as (42) but cannot be destroyed (43).

$$\int_0^L P_{Ni,(x,y,z)} dr_s = P_{NL} + P_{NR}$$

(42)

$$\int_0^L P_{Ni,(x,y,z)} dr_s \neq P_{NL} + P_{NR}$$

(43)

Some experiments are proposed to answer the following questions. First, can the regeneration as two new *space waves* χ_{PL} and χ_{PR} be proved? In single particle double slit experiments, the particle's path is midway between the slits. Shifting this position sideways, with respect to the slits, should demonstrate a phase shifted interference pattern that is governed by this shift.

Second, can the interference *envelope function* φ, exists behind the particle? The direction of motion and the probability of localization need to be consistent with each other. Therefore, this answer should be, that it cannot. A two particle double slit experiment with the screen behind the slits should not show any interference patterns. If interference patterns are not observed behind the slits, it proves that interference probabilities are asymmetrical and do not occur in the same magnitude in every direction.

Third, does the interference *space wave* χ exist behind the aperture? That is, is the *space wave* χ symmetric about the *x-y* plane in the *z*-direction? A modified two particle double slit experiment with a photon barrier to prevent *self-interference*, could answer this question. See Fig. 5. Instead of a pair photons travelling through their respective slits in the same direction, one photon travels in the opposite direction. If the *space wave* χ is *x-y* symmetric then one should observe interference patterns on both screens, in front and behind. If it is, then $\psi=1$ along the -z axis.

Fourth, can *space wave* χ independence with respect to particle mass be proved? Again, the double slit experiment can confirm or deny this. In the double slit experiment, let one particle be a photon and the other an electron. Since both interference *space waves* χ are of the same type and in the same medium of spacetime, interference patterns should be observed, and would be substantial proof of this mass independence. Thereby, increasing the possibility that the Schrödinger wave function Ψ_S could be replaced by the *Probabilistic Wave Function* Ψ_P, its solution ψ_P, and components, *space wave* χ_P and *envelope function* φ_P.

Since the particle's axis of propagation is between the double slits, the inference is that the particle does not travel through the barrier between slits because the slits exists. The center photon barrier in Fig. 5 should confirm this. Thus, a *barrier* is defined as an obstruction that prevents the formation of the *space wave* χ (not the particle's χ_P).

Therefore, the inference is that a particle exists in an *unlocalized form* and its *space wave* χ_P determines its most likely position. Since it is *unlocalized*, its *envelope function* φ_P determines where it will localize. And thus, localization could mean coming out, into spacetime, but this requires an answer, where from? Solomon [1], [4], [5], [6] & [7] proposed different types of spacetime, termed *kenos* (from Greek for vacuous), coexists with spacetime. One of these is *subspace* as a *kenos (x, y, z)* without the time dimension, and on localization, particles emerge from the *subspace kenos*. For nuclei structure Solomon [1] proposed another type of *kenos*, *Variable Electric Permittivity (VEP)* matter.

5.7 Quantum Entanglement

Locality demands the conservation of causality, meaning that information cannot be exchanged between two space-like separated parties or actions [16]. Quantum entanglement can be described [17] as non-local interactions or the idea that distant particles do interact without hidden variables.

There are two possible alternative explanations that do not require hidden variables. First, that subspace is the carrier of this entangled information, and second is the *envelope function* φ_P based spatial probability distri-

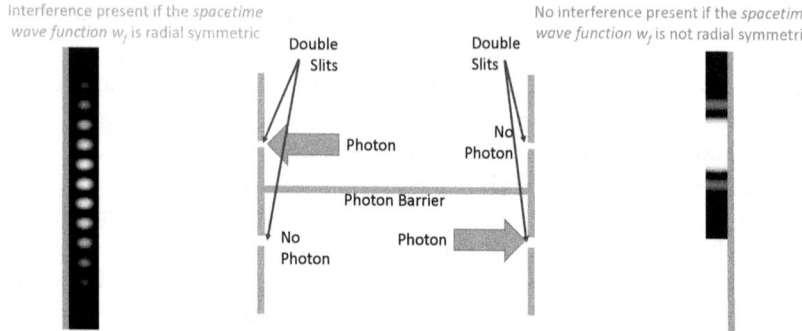

Fig. 5, Test for Radial Symmetric spacetime term χ

bution. The spatial probability distribution is so large and long [8] that entanglement occurs while the entangled photons' probability fields overlap. The joint probability $P_{ij,(x,y,z)}$ (44) at coordinates (x, y, z) of photon i interacting with its entangled photon j is the product of the individual probabilities $P_{i,(x,y,z)}$ and $P_{j,(x,y,z)}$.

$$P_{ij,(x,y,z)} = P_{i,(x,y,z)}P_{j,(x,y,z)} \qquad (44)$$

Given that these particle probabilities obey the *envelope function* φ_P (25), (44) can be written as (45),

$$\varphi_{Pij,(x,y,z)} = \varphi_{Pi,(x,y,z)}\varphi_{Pj,(x,y,z)} \qquad (45)$$

The normalized probabilities P_N would take the form,

$$P_{N,ij,(x,y,z)} = P_{N,i,(x,y,z)}P_{Nj,(x,y,z)} \qquad (46)$$

with,

$$P_{Ni,(x,y,z)} = \varphi_{Pi,(x,y,z)}/F_{I,A'} \qquad (47)$$

$$P_{Nj,(x,y,z)} = \varphi_{Pj,(x,y,z)}/F_{I,A'} \qquad (48)$$

Assuming that the particle must arrive at the detector, the denominator is the *field of interaction* $F_{I,A'}$. This is the area of the curved surface of the detectors used in the experiments. This denominator can be simplified to $F_{I,A}$ by only considering photon arrivals through the cross section of the detector aperture as opposed to detector surface.

If entanglement is due to the joint probabilities of the *envelope function* φ_P then two scenarios can be formulated:

i) Two un-entangled independent photons i & j with their *envelope terms* φ_{Pi} & φ_{Pj} pass through their respective double slits, will exhibit the normal interference patterns. Mathematically, by linear superposition, these are the straight sum $\Sigma_n \psi_{Pn}$ of their respective, *probabilistic wave solution* ψ_{Pn}, for $n=i$ & j, and (16) would hold, such that (Fig. 6),

$$\sum_n \psi_{Pn} = \varphi_{Pi}\chi_{Pi} + \varphi_{Pj}\chi_{Pj} \tag{49}$$

ii) If two entangled photons i & j exhibit joint probabilities, then their *envelope terms* φ_{Pi} & φ_{Pj} are replaced by their *joint probability envelop term* φ_{PJ}, (36), such that the joint *probabilistic wave solution* ψ_{PJn} by linear superposition and (16), would take the form (Fig. 7),

$$\sum_n \psi_{PJn} = \varphi_{PJ}\chi_{Pi} + \varphi_{PJ}\chi_{Pj} \tag{50}$$

For ease of illustration, Fig. 6 & 7, the photons i & j are separated by *1m* with the *probabilistic wave functions* mapped out on a *10m* square screen, positioned *1m* away from their apertures.

It is clear that the resulting interference patterns are different and therefore a test of the validity of subspace versus joint probability. If subspace is the carrier of entangled information one should observe interference patterns similar to Fig. 6, else if joint probabilities are the carrier than one should observe interference patterns similar to Fig. 7.

5.8 Electron Shell

The ultimate test of any new theoretical approach to the foundations of physics must be the electron shell. This paper proposes the *Orbital Space Wave (OSW)* electron shell model that in essence electron orbitals are at least half integer numbers of the nucleus' *space wave* χ_P periods i.e. at its nodes. To arrive at this model several factors need to be accounted for. Given that electric field lines are not repulsive and thus, Poincaré stresses [21] do not exist [2], the nuclear electric field permeates throughout the electron shell. Therefore, the shrinking size of the electron shell cannot be due to a shielding effect if it no longer exists.

Since the electric field force is to the power of -2 of distance this would suggest that the closer electrons would negate the nuclear attractive force with a net repulsive. A simple model of the net forces F_N on the last electron in an electron shell of n electrons is,

$$F_N = \frac{nQQ}{4\pi\varepsilon_0 r^2} - \frac{(n-1)QQ}{4\pi\varepsilon_0 \delta r^2} \tag{51}$$

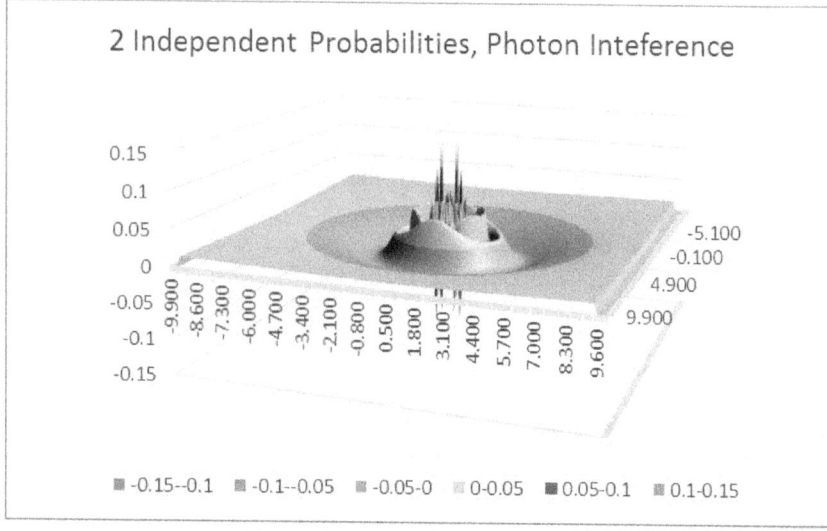

Fig. 6, Airy disc interference of 2 photons with independent envelop probabilities

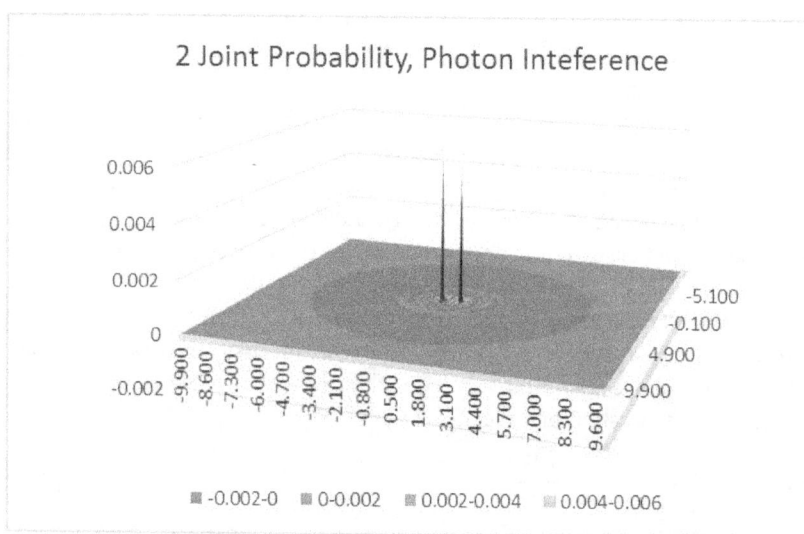

Fig. 7, Airy disc interference of 2 photons with joint envelope probabilities

Where r is the distance of the last electron from the nucleus and δr is the average distance of the n-1 electrons from the last. The boundary condition for attraction is that

$$F_N \geq 0 \tag{52}$$

or

$$\delta r^2 \geq r^2 \frac{(n-1)}{n} \tag{53}$$

That is, as n becomes large, the average distance δr from the remaining electrons approaches the distance of the last electron from the nucleus, which is not possible. One could build more sophisticated models but the conclusion would be similar. It shows that this electric field approach to contracting electron shell does lead to issues of stability as these electrons 'orbit' the nucleus.

Using Slater's data [20, page 22] a regression fit of the Z number of protons with the atomic radius (for Li, Be, B, C, N, & O) shows that the best fit (R^2=99.25%) relationship is an inverse function of Z. That is, the net electron shell electric forces are an inverse function of charge. This is *correlation* and not *logically* correct as electric forces are directly proportional to charge.

The other possibility is that Z is a proxy for some other factor that is related to the number of protons. That factor would be mass. de Broglie's equation shows that the rest mass wavelength λ_0 shrinks as mass increases, a property that could explain electron shell shrinkage.

The pre-Quantum Mechanical Rydberg[ix] equation (46) would be good starting point.

$$\frac{1}{\lambda_{f,i}} = R_H \left(\frac{1}{n_f^2} - \frac{1}{n_i^2} \right) \tag{54}$$

Where n_i is the initial orbital number, n_f is the final orbital number, and R_H is the Rydberg constant. Four modifications to this Rydberg equation (46) are required.

i) Include the de Broglie rest mass wavelength λ_B for the proton masses. Neutron masses cannot be included as these are not related to the Z numbers.

ii) This then enables the electron transitions to be described in terms of the *space waves* χ_P.

iii) To be consistent across all electron shells the Hydrogen H electron shell properties should determine the properties of the other elements, as Bohórquez & Boyd [20, page 16] had suggested

Table 1, Orbital Space Wave Model for Hydrogen, in terms of number of λ_H periods.

$n_{\gamma L} \equiv$	$n_{\gamma L}$	$n_{\gamma U} \mid n_{\gamma L} \equiv 2$	$n_{\gamma U} \mid n_{\gamma L} \equiv 3$
1	1.201199E+15		
2	3.175175E+15		
3	4.941522E+15	3.651429E+15	
4	6.660546E+15	3.838409E+15	5.187802E+15
5	8.354003E+15	3.929367E+15	5.306253E+15
6	1.002700E+16	3.980057E+15	5.371882E+15
7	1.168030E+16	4.011078E+15	5.411911E+15
8	1.331320E+16	4.031401E+15	5.438079E+15
9	1.492438E+16	4.045422E+15	5.456107E+15
10	1.651236E+16	4.055495E+15	5.469047E+15
11	1.807559E+16	4.062972E+15	5.478644E+15
12	1.961260E+16	4.068672E+15	5.485957E+15
13	2.112200E+16	4.073117E+15	5.491657E+15
14	2.260252E+16	4.076649E+15	5.496184E+15
15	2.405302E+16	4.079501E+15	5.499840E+15
16	2.547255E+16	4.081838E+15	5.502834E+15
17	2.686029E+16	4.083776E+15	5.505317E+15
18	2.821554E+16	4.085401E+15	5.507399E+15
19	2.953779E+16	4.086778E+15	5.509162E+15
20	3.082664E+16	4.087953E+15	5.510667E+15
21	3.208187E+16	4.088965E+15	5.511963E+15
22	3.330331E+16	4.089843E+15	5.513087E+15
23	3.449100E+16	4.090609E+15	5.514067E+15
24	3.564502E+16	4.091281E+15	5.514928E+15
25	3.676557E+16	4.091874E+15	5.515687E+15
26	3.785296E+16	4.092401E+15	5.516361E+15
27	3.890755E+16	4.092870E+15	5.516961E+15
28	3.992981E+16	4.093289E+15	5.517499E+15
29	4.092024E+16	4.093666E+15	5.517981E+15
30	4.187942E+16	4.094007E+15	5.518417E+15
31	4.280794E+16	4.094314E+15	5.518811E+15
32	4.370650E+16	4.094594E+15	5.519168E+15
33	4.457576E+16	4.094848E+15	5.519494E+15
34	4.541644E+16	4.095080E+15	5.519791E+15
35	4.622928E+16	4.095293E+15	5.520063E+15
36	4.701502E+16	4.095488E+15	5.520313E+15
37	4.777445E+16	4.095668E+15	5.520543E+15
38	4.850829E+16	4.095834E+15	5.520755E+15
39		4.095987E+15	5.520951E+15
40		4.096128E+15	5.521132E+15

$$r_a = a_0 \sqrt{I_H/I} \tag{55}$$

iv) Where r_a is the atomic radius, a_0 is the Bohr radius and I_H and I are the ionization potentials of Hydrogen and the element in question, respectively.

v) Additionally, the orbital numbers n_f and n_i should provide complete information about the state of the electrons as these are proxies for distances from the nucleus and one would not need to take into account the quantum mechanical angular momentums, s, p, d, f, ... of these electrons. These angular momentums $1s$, $2s$, $2p$, ..., are replaced by orbital numbers 1, 2, 3, ..., respectively.

Building on these four constraints, using National Institute of Standards and Technology (NIST)[x] spectral database, extensive numerical modeling of this spectral data for 5 elements, from Hydrogen H to Boron B, shows that the spectral wavelength $\lambda_{L,U}$ derived from the electron transition from lower $n_{\chi L}$ to the upper $n_{\chi U}$ integer numbers of periods of the element's *space wave* χ, wavelength λ_E is of the form,

$$\frac{1}{\lambda_{L,U}} = R_H \left[\left(\frac{\lambda_H}{\lambda_E} \right)^2 n_\chi \right] \left(\frac{1}{n_{\chi L}^2} - \frac{1}{n_{\chi U}^2} \right) + e_j \tag{56}$$

Where λ_H is the de Broglie rest mass wavelength of the Hydrogen proton mass (AMU, kg), λ_E is the de Broglie rest mass wavelength of element E's proton (AMU, kg) masses, e_j is the spectral error introduced by the quantum mechanical *term* & *J* components and,

$$n_\chi = \left(\frac{1}{n_{\chi L}^2} + \frac{1}{n_{\chi U}^2} \right) \tag{57}$$

Simplifying, gives,

$$\frac{1}{\lambda_{L,U}} = R_H \left(\frac{\lambda_H}{\lambda_E} \right)^2 \left(\frac{1}{n_{\chi L}^4} - \frac{1}{n_{\chi U}^4} \right) + e_j \tag{58}$$

Table 1, provides the values of $n_{\chi L}$ and $n_{\chi U}$ for 3 of the 15 partial columns for Hydrogen.

The error for the Rydberg equation (54) is about $\pm 0.02\%$ for Hydrogen. The *Orbital Space Wave* (56) model for the 5 elements, H, He, Li, Be & B was constructed using Hydrogen's number of periods $n_{\chi L}$ in column 1 of Table 1 as the lower number of periods $n_{\chi L}$ for these elements. The upper levels $n_{\chi U}$ vary with each element.

Comparing quantum mechanics' Ritz wavelength of the NIST spectral database with *Orbital Space Wave* (56) model gives an error $\leq \pm 1 \times 10^{-9}\%$ at most or identical spectral frequencies with quantum mechanics. This confirms the thesis that electron distance from the nucleus, in terms of *space wave* χ_P periods *is sufficient* to determine spectral frequencies and therefore, electron shell arrangement.

Even though the *Orbital Space Wave* (56) model can be described as a success, there still is much work to be done to get it to where quantum mechanics is, such as refining how these orbitals are structured, a physical interpretation, a model for chemical reactions, how does the electron's *envelope terms* φ_P interact with these orbitals, and much more.

5.9 Conclusion

There is a replacement for the Schrödinger wave function Ψ_S is the *Probabilistic Wave Function* Ψ_P. Unlike Schrödinger's Ψ_S whose wave packet is a damped oscillating standing wave, the *Probabilistic Wave Function* Ψ_P wave packet is an inverse function of distance. The proposed *Probabilistic Wave Function* Ψ_P and its components *space wave* χ_P and *envelope function* φ_P explain how interference, entanglement and electron shell occur. More research is required to explore how *Probabilistic Wave Function* Ψ_P can explain other quantum mechanical phenomena.

References:

[1] Solomon, B.T. "A Non Standard Model Nucleon/Nuclei Structure",, in *Super Physics for Super Technologies*, March 2015.

[2] Solomon, B. T., "The Variable Isotopic Gravitational Constant", in *Super Physics for Super Technologies*, March 2015..

[3] Solomon, B.T., "A Universal Approach to Forces", in *Super Physics for Super Technologies*, March 2015..

[4] Solomon, B. T., "Empirical Evidence Suggest A Need For A Different Gravitational Theory," in the proceedings of the *100 Year Starship Study Public Symposium (100YSS,)*, 2013.

[5] Solomon B.T., "New Evidence, Conditions, Instruments & Experiments for Gravitational Theories", Journal of Modern Physics, Special Issue on Gravitation, Astrophysics and Cosmology, Vol. 8A, 2013, August 2013.

[6] Solomon, B. T., "Empirical Evidence Suggest A Need For A Different Gravitational Theory," *American Physical Society (APS) April Conference, Denver,* 2013

[7] Solomon B.T., *An Introduction to Gravity Modification: A guide to using Laithwaite's and Podkletnov's experiments and the physics of forces for empirical results, 2nd Edition.* Universal Publishers, Boca Raton, May 2012.

[8] Solomon, B. T., "Non-Gaussian Radiation Shielding," *100 Year Starship Study Public Symposium (100YSS,),* 2011.

[9] Solomon, B.T., "Gravitational Acceleration Without Mass And Noninertia Fields", Physics Essays, Vol. 24, 327, 2011. [Phys. Essays **24**, 327 (2011)]

[10] Solomon, B. T., "Reverse Engineering Podkletnov's Experiments," in the proceedings of the *Space, Propulsion & Energy Sciences International Forum (SPESIF-11),* Edited by Glen A Robertson, Physics Procedia, Elsevier Science.

[11] Solomon, B. T., "Non-Gaussian Photon Probability Distributions," in the proceedings of the *Space, Propulsion & Energy Sciences International Forum (SPESIF-10),* Edited by Glen A Robertson, AIP Conference Proceedings **1208**, Melville, New York, (2010).

[12] Solomon, B. T., "An Approach to Gravity Modification as a Propulsion Technology," in the proceedings of the *Space, Propulsion & Energy Sciences International Forum (SPESIF-09),* Edited by Glen A Robertson, AIP Conference Proceedings **1103**, Melville, New York, (2009).

[13] Solomon, B. T., "An Epiphany On Gravity", Journal of Theoretics, Vol. 3-6, 2001.

[14] Feynman, R., *Feynman Lectures on Physics,* Chapter 28-4.

[15] Nemiroff, R., "Bounds on Spectral Dispersion from Fermi-Detected Gamma Ray Bursts," Physical Review Letters, Vol. 108, No. 23, 2012.

[16] Eisaman, M. D., Goldschmidt, E. A., Chen, J., Fan, J. and Migdall, A., "Experimental test of nonlocal realism using a fiber-based source of polarization-entangled photon pairs," *Phys. Rev. A,* **77**, (2008), p. 032339.

[17] Howell, J. C., Bennink, R. S., Bentley, S. J. and Boyd, R. W., "Realization of the Einstein-Podolsky-Rosen Paradox Using Momen-

tum and Position-Entangled Photons from Spontaneous Parametric Down Conversion," *Phys. Rev. Lett.*, **92**(21), (2004), pp. 210403-1-4.

[18] Angeli, I., "A consistent set of nuclear rms charge radii: properties of the radius surface (R,N)", Science Direct, Volume 87, Issue 2, July 2004, Pages 185–206, DOI: 10.1016/j.adt.2004.04.002

End Notes:

i. Does not include non-peer reviewed conference presentations between 2001 and 2009.

ii. The particle is traveling along the z-axis, with the orthogonal screen in the x-y plane.

iii. Macro body elongation due to tidal gravity is attributed to Roger Penrose (this author could not find the reference in time for this paper). Macro bodies elongate as the body falls into a gravitational field. Let's reexamine this tidal behavior with the additional requirement that this tidal gravity property be consistent with Lorentz-FitzGerald transformations or Special Theory of Relativity. To be consistent with Lorentz-FitzGerald transformations, the atoms and elementary particles would contract in the direction of the fall. However, to be consistent with tidal gravity's elongation, the distances between atoms in the macro body has to increase at a rate consistent with the acceleration and velocities experienced by the various parts of the macro body. That is, as the atoms get flatter, the distances apart get longer. One suspects that this axiom's inconsistency with the empirical evidence has led to an explosion of string theories, each trying to explain Nature with no joy.

iv. http://www.colorado.edu/physics/TZD/PageProofs1/TAYL07-203-247.I.pdf

v. http://hyperphysics.phy-astr.gsu.edu/hbase/oscda.html

vi. Even though the exponential term +1/u has a positive sign compared to the negative in –γ, the u function is an inverse.

vii. Regression fitting the data (1/u against $\exp^{+1/u}$) gives R^2 of 99.944%, 99.562%, 98.131%, 95.846% & 92.959% at screen distances, d_s of 0.10m, 0.25m, 0.50m, 0.75m & 1.00m, respectively. Compared to an exact theoretical model producing an R^2 of 99.9999%. That is the damped oscillation model fails as d_s increases. This implies that a different model is at work.

viii. http://hyperphysics.phy-astr.gsu.edu/hbase/waves/wpack.html

ix. http://chemwiki.ucdavis.edu/Physical_Chemistry/Quantum_Mechanics/Quan-
tum_States_of_Atoms_and_Molecules/2._Foundations_of_Quantum_Mechanics/Derivation_of_the_Rydberg_Equation_from_Bohr's_Model

x. http://physics.nist.gov/PhysRefData/ASD/lines_form.html

xi. For these elements, a total of 7,972 spectral lines were reduced to 1,963 by removing duplicates and incompletes. For duplicate data the average spectral wavelength was computed. Data with problems or errors were removed. This investigation was limited to the *s, p, d, f equivalent numbering* of $1 \leq n_{\chi L} \leq 15$ and stated in integer number of Hydrogen's wavelength λ_H, and $2 \leq n_{\chi U} \leq 100$ stated in integer number of the element E's wavelength λ_{E}.

6. Particle Structure

Abstract:- This paper proposes that there is a hierarchy to particle design and, both mass and massless particles can be built from individual components consisting of the *scalar proto-fields, null direction lines, rings, discs* and *spheres*. Rings, discs and spheres are how particles exhibit quantized properties.

Combined with the wave function these *proto-fields* take on positive or negative values to form their respective fields. This approach explains why some properties reverse in the anti-particle, and some do not. To build on this requires a building block approach proposed in this paper.

It is proposed that mass is an electromagnetic *ring*, and therefore does not change its sign in the anti-particle as the sign change is evidenced as a phase change.

Ring fields explain spin, Pauli's Exclusion Principle, and magnetic moments, while spheres explain electric and magnetic monopole charges.

It is proposed that the electromagnetic transverse wave in the photon is derived from the *proto-field null direction lines*. This energy is the rotation of these vectors between spacetime and subspace. Thus, particles only carry energy and mass by their rotating electromagnetic vectors.

A *Component Standard Model* is proposed that shows how the neutron decay works and why quarks are apparently confined. Finally, a more detailed explanation of what photon localization is, is proposed.

6.1 Introduction

This paper reports the 15th paper in the 16-year ([1] to [14] [i]) investigation into the feasibility of gravity modification, and a *New Standard Model*.

Primarily, that the spatial gradients of fields is a universal mechanism by which forces are transmitted [4], the Schrödinger wave function [1] has a replacement (1), Poincaré stresses [15] do not exist [4], quantum foam does not exists [16], the gravitational constant G is a variable [3] and there exists a massless formula for gravitational acceleration g (5),

$$\psi_P = \varphi_P \chi_P = \left(1/k_\psi x\right)\sin\left(k_\psi x\right) \qquad (1)$$

Where the *space wave* χ_P is,

$$\chi_P = \sin\left(k_\psi x\right) \qquad (2)$$

And the *envelope function* φ_P is

$$\varphi_P = \left(1/k_\psi x\right) \qquad (3)$$

Where x is the radial distance from the photon axis of propagation.

$$k_\psi = (2\pi/\lambda) \qquad (4)$$

$$g = \tau c^2 \qquad (5)$$

where τ is the change in time dilation divided by the change in that distance.

These findings should lead to the replacement of the Standard Model with something substantially simpler. Occam's Razor, that the simpler solution is most likely the correct solution, affirms this investigation.

6.2 Modular Building Blocks

This paper proposes that all particle properties are constructed from particle components or *scalar proto-fields* in modular building blocks. There are four types of *scalar proto-fields*, *null direction lines* ι, *rings* ϱ, *discs* υ, and *spheres* ξ. An example of an electric *null direction line* ι_E is an electric field vector without the vector direction. See section 4 for rings. The *space wave* χ_P and the *envelope function* φ_P are radial descriptions of these *disc* υ *proto-fields*. Note that unlike *lines* ι and *discs* υ, *rings* ϱ, and *spheres* ξ are hollow in the center.

A *proto-field* acquires it full potential of that of a field, in the presence of the wave function. For example, electric *null direction line* ι_E is an electric field vector $+E$ when it occupies the *space wave* χ_P (2) at $x=\lambda/4$, and $-E$ at $x=3\lambda/4$. Similarly, electric charge is positive or negative if the charge's *spherical* ξ_Q *proto-field* has a radius of $\lambda/4$ or $3\lambda/4$, respectively. It is then possible to design any type of particle from these building blocks.

A particle's building block consists of field building blocks, F_B (6) that can be structured as consisting of two field components, the *field attenuation* F_A and *field derivative* F_D.

$$F_B = F_A F_D \qquad (6)$$

The *field derivative* F_D itself is composed of the *field origin* F_O which is a function the specific particle property and not a function of either space or time.

The first of these building blocks is the particle's [14] *Probabilistic Wave Function* Ψ_P which is also the *probabilistic wave solution* ψ_P (1) as the solution is non integrable. Thus,

$$F_B = \psi_P = F_A F_D = \varphi_P \chi_P \qquad (7)$$

Such that,

$$F_{A,\psi} = \varphi_P = \left(1/k_{B,\psi} x\right) \qquad (8)$$

$$F_{D,\psi} = \chi_P = sin\left(k_{B,\psi} x\right) \qquad (9)$$

Where,

$$k_{B,\psi} = (2\pi/\lambda)\iota_\psi = (2\pi/\lambda)(+1) \qquad (10)$$

Where ι_ψ is the dimensionless *probabilistic null direction line proto-field*, $k_{B,\psi}$ is the *probabilistic field attenuation constant* of the *field attenuation* $F_{A,\psi}$, and the *field origin* $F_{O,\psi}$ is,

$$F_{O,\psi} = k_{B,\psi} = (2\pi/\lambda)\iota_\psi \qquad (11)$$

That is, in this case the attenuation constant is also the field origin. Note that (8) and (9) are dimensionless which should be the case for a probabilistic function.

Particle design note #1, that the *field origin* F_O is independent of space or time.

The second building block is the electric field. Using the nomenclature of building blocks, a charged $Q_{P,E}$ particle's *electric field* E_P at any radial distance r from its center with an *electric permittivity* ε, gives,

$$F_{B,E} = E_P = (1/4\pi\varepsilon r)\left(Q_{P,E}/r\right) \qquad (12)$$

Such that,

$$F_{A,E} = \left(k_{B,E}/r\right) \qquad (13)$$

$$F_{D,E} = \left(Q_{P,E}/r\right) \qquad (14)$$

Where,

$$k_{B,E} = 1/4\pi\varepsilon \qquad (15)$$

Where $k_{B,E}$ is the *electric field attenuation constant* of the *field attenuation* $F_{A,E}$ and the *field origin* $F_{O,E}$ is,

$$F_{O,E} = Q_{P,E} = sin\left(\pm\frac{\pi}{2}\right)n\xi_{P,E} \tag{16}$$

Where n is the number of *electric proto-field spheres* $\xi_{P,E}$ ($= \frac{1}{3}\, e$, electric charge), in the charge. Similarly, the third building block is the magnetic field. It is that of a magnetic monopole charge $Q_{P,M}$ particle's *magnetic field* B_P at any radial distance r from its center with *magnetic permeability* μ, gives[ii],

$$F_{B,M} = B_P = (\mu/4\pi r)(Q_{P,M}/r) \tag{17}$$

Such that,

$$F_{A,M} = (k_{B,M}/r) \tag{18}$$

$$F_{D,M} = (Q_{P,M}/r) \tag{19}$$

Where,

$$k_{B,M} = \mu/4\pi \tag{20}$$

Where $k_{B,M}$ is the *magnetic field attenuation constant* of the *field attenuation* $F_{A,M}$ and the *field origin* $F_{O,M}$ is,

$$F_{O,M} = Q_{P,M} = sin\left(\pm\frac{\pi}{2}\right)n\xi_{P,M} \tag{21}$$

Where n is the number of *magnetic proto-field spheres* $\xi_{P,M}$, in the magnetic monopole. Particle design note #2, that the *field attenuation* F_A and *field derivative* F_D, are spatial functions.

6.3 Transverse EM Wave

The electric field building block can be adapted to the transverse electromagnetic (TEM) wave. The TEM electric field can be represented by its *null direction line proto-field* ι_E, and a single field line of maximum amplitude E_A, a constant at sin(90°), in the x-y plane orthogonal to the direction of propagation, the z-axis. From (13) its *field attenuation* $F_{A,E}$ is,

$$F_{A,E} = (k_{B,E}/r) = E_A = sin\left(\pm\frac{\pi}{2}\right)n\iota_E \tag{22}$$

Assuming $n=1$, (22) shows that the value of ι_E is E_A. Therefore, the *field derivative* $F_{D,E}$ (23) of this single field line, is interpreted as the electric charge per unit length $Q_{P,E}/r$. It is also a constant as by (22) r is no longer a variable. Thus,

$$F_{D,E} = Q_{P,E}/r = 4\pi\varepsilon E_A Q_{P,E} \tag{23}$$

Now, for an oscillating electric field $F_{B,EW}$, with field strength E_A and wavelength λ propagating along the z-axis, by (6),

$$F_{B,EW} = F_{A,EW}F_{D,EW} \tag{24}$$

$$F_{B,EW} = E_A sin((2\pi/\lambda)z) = E_A sin(k_{B,\psi}z) \tag{25}$$

By particle design note #1 E_A is the *field derivative* $F_{D,EW}$ as it contains the *field origin* $F_{O,E}$,

$$F_{D,EW} = E_A \tag{26}$$

the *field attenuation* $F_{A,EW}$ along the z-axis is given by,

$$F_{A,EW} = sin(k_{B,\psi}z) \tag{27}$$

Or the *Probabilistic Wave Function* Ψ_P, is the origin of the *electric field attenuation* $F_{A,EW}$,

Similarly, for the magnetic field, this TEM magnetic field can be represented by its *null direction line proto-field* ι_M, and a single field line of maximum amplitude B_A, a constant at sin(90°), in the x-y plane orthogonal to the direction of propagation, the z-axis. From (18) its *field attenuation* $F_{A,M}$,

$$F_{A,M} = (k_{B,M}/r) = B_A = sin\left(\pm\frac{\pi}{2}\right)n\iota_M \tag{28}$$

Assuming *n=1*, (28) shows that the value of ι_M is B_A. Therefore, the *field derivative* $F_{D,M}$ (29) of this single field line, is interpreted as the magnetic charge per unit length $Q_{P,M}/r$. It is also a constant as by (28) r is no longer a variable. Thus,

$$F_{D,M} = Q_{P,M}/r = \frac{4\pi}{\mu}B_A Q_{P,M} \tag{29}$$

Now, for an oscillating magnetic field $F_{B,MW}$, with field strength B_A and wavelength λ propagating along the z-axis, by (6),

$$F_{B,MW} = F_{A,MW}F_{D,MW} \tag{30}$$

$$F_{B,MW} = B_A sin((2\pi/\lambda)z) = B_A sin(k_{B,\psi}z) \tag{31}$$

By particle design note #1 B_A is the *field derivative* $F_{D,MW}$ as it contains the *field origin* $F_{O,M}$,

$$F_{D,MW} = B_A \tag{32}$$

the *field attenuation* along the z-axis is given by,

$$F_{A,MW} = sin(k_{B,\psi}z) \tag{33}$$

Or the *Probabilistic Wave Function* Ψ_P, is the origin of the *magnetic field attenuation* $F_{A,MW}$, and therefore, the *Probabilistic Wave Function* Ψ_P is the origin of the electromagnetic wave.

Checking the conservation of mass-energy shows that one more building block is required. The electric η_E and magnetic η_M field energy densities[iii] are given by,

$$\eta_E = \left(\frac{1}{2}\right)\left(\varepsilon F_{B,EW}^2\right) = \left(\frac{\varepsilon}{2}\right)\left(E_A sin\left(k_{B,\psi}z\right)\right)^2 \tag{34}$$

$$\eta_M = \left(\frac{1}{2}\right)\left(\frac{F_{B,EW}^2}{\mu}\right) = \left(\frac{1}{2\mu}\right)\left(B_A sin\left(k_{B,\psi}z\right)\right)^2 \tag{35}$$

That is, the energy densities oscillates between 0 and maximum. Solomon proposed [2], [3] & [4] the existence of multiple types of *kenos* (Greek for vacuous) of spacetime. One being subspace (x, y, z) which does not have the time dimension. An interaction occurs between *spacetime kenos a* and *subspace kenos β* when they are joined in that local region of space. Thus probabilistic behavior occurs at the particle level because the *join* is at the particle level.

Thus, as the transverse wave's electric (27) and magnetic (33) field oscillations are governed by the *Probabilistic Wave Function* Ψ_P, these oscillations occur in the *joined kenos set* $\{a, \beta\}$ and rotate at $k_{B,\psi}$ $(=2\pi/\lambda)$ between *spacetime kenos a* and *subspace kenos β*. For conservation of mass-energy to hold these *kenoses* are 90° out of phase, or

$$_\alpha E_A = E_A sin\left(k_{B,\psi}\left(_\alpha z\right)\right) \tag{36}$$

$$_\beta E_A = E_A sin\left(k_{B,\psi}\left(_\beta z\right)+\frac{\pi}{2}\right) = E_A cos\left(k_{B,\psi}\left(_\beta z\right)\right) \tag{37}$$

$$_\alpha B_A = B_A sin\left(k_{B,\psi}\left(_\alpha z\right)\right) \tag{38}$$

$$_\beta B_A = B_A sin\left(k_{B,\psi}\left(_\beta z\right)+\frac{\pi}{2}\right) = B_A cos\left(k_{B,\psi}\left(_\beta z\right)\right) \tag{39}$$

And total electrical energy $_T\eta_E$ and total magnetic energy $_T\eta_M$ are the sum of their respective components in the *spacetime kenos a* and *subspace kenos β*, such that these total energies are always a constant,

$$_T\eta_E = {_\alpha\eta_E} + {_\beta\eta_E} = \left(\frac{\varepsilon E_A^2}{2}\right) \tag{40}$$

Where,

$$_\alpha\eta_E = \left(\frac{\varepsilon E_A^2}{2}\right)sin\left(k_{B,\psi\ \alpha}z\right)^2 \tag{41}$$

$$_\beta\eta_E = \left(\frac{\varepsilon E_A^2}{2}\right)cos\left(k_{B,\psi\ \beta}z\right)^2 \tag{42}$$

And,

$$_T\eta_M = {_\alpha\eta_M} + {_\beta\eta_M} = \left(\frac{B_A^2}{2\mu}\right) \tag{43}$$

Where,

$$_\alpha\eta_M = \left(\frac{B_A^2}{2\mu}\right) sin\left(k_{B,\psi}\ _\alpha z\right)^2 \tag{44}$$

$$_\beta\eta_M = \left(\frac{B_A^2}{2\mu}\right) cos\left(k_{B,\psi}\ _\beta z\right)^2 \tag{45}$$

Such that, both the total electric η_E and total magnetic η_M field energies are constant. That is, as these fields oscillate between *spacetime kenos* α and *subspace kenos* β, their energies do so, too.

Should the phase change, this would be equivalent to raising the wave in one *kenos* and lowering it in the other. However, this phase change is not allowed as it would break conservation of mass-energy.

One infers that when the electric ι_E and magnetic ι_M *null direction line proto-fields* switch *sin(±π/2)* between *spacetime kenos* α and *subspace kenos* β, the *proto-fields* locations on the *Probabilistic Wave Function* Ψ_P also switch between *+1* & *-1*.

Therefore, the fourth building block is the *kenos* or *spacetime* α and *subspace* β, *kenoses*. Note that unlike imaginary *i* numbers β cannot be negative. A negative would be a property of the field being described, not the *kenos*. The α and β nomenclature provide a mechanism to identifying where what is.

6.4 Ring Fields

Quantum mechanics requires that magnetic moment μ_{spin} is a function of the particle's charge, *e per* (46),

$$\mu_{spin} = \frac{e}{m_e} s_z \tag{46}$$

However, the ratio of the proton to neutron magnetic moments is - 3/2[iv]. If one were to ignore the sign of the quark charges in theses nucleons, the magnitude of these quark charges Q_q, would range as follows,

$$0 \le Q_q \le \frac{4}{3}e \tag{47}$$

Even taking into account the nucleon masses, one would expect a maximum ratio of 4/3, but this is not the case. It is 8.8% bigger at -1.461910. This would suggest that a particle's magnetic moment is not related to its electric charge, but to something else that is equivalent to charge. May be not at all. Therefore, the *ring field* ϱ_F proposal.

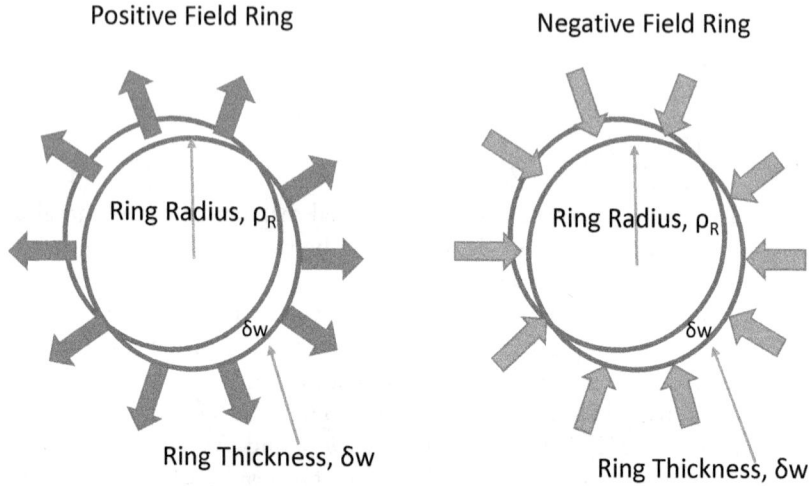

Fig. 1, Positive and Negative Electric Ring Fields with radius ϱ_R

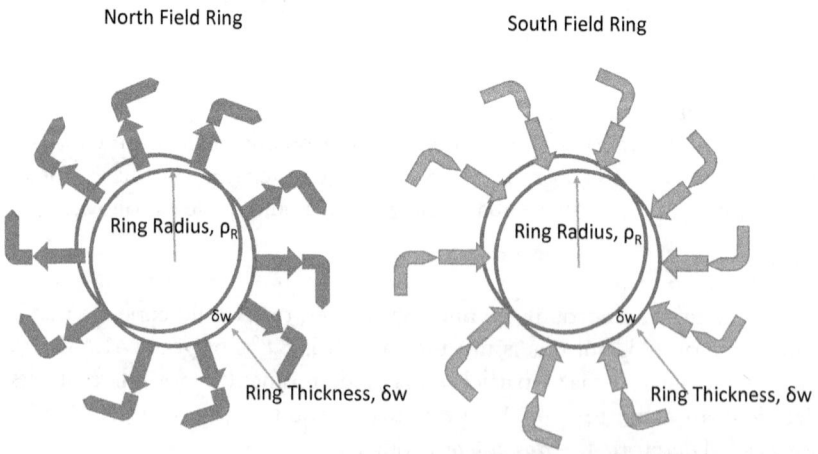

Fig. 2, North and South Magnetic Ring Fields with radius ϱ_R

Particle Structure

The fifth building block is the *ring field* ϱ_F as illustrated in Fig. 1. This paper proposes, particle design note #3, that any field *type F* can be derived by any type of *proto-field*. For example, the *ring fields probabilistic* ϱ_ψ, *electric* ϱ_E, and *magnetic* ϱ_M. Further, the strength of the *ring field* ϱ_F is much less than the other types of *proto-fields*. The rings have field lines entering (green arrows) or exiting (red arrows) the ring. The base of the ring (blue) is the end or beginning of these field lines, and forms a cylinder of radius ϱ_R with a very narrow width δw. In essence the *ring field* ϱ_F is equivalent to a slice of the particle's spherical electric or magnetic monopole fields such that its thickness δw,

$$\delta w \approx 0 \tag{48}$$

and

$$\delta w > 0 \tag{49}$$

Using the convention of electric fields, Fig.1 shows the *electric ring* ϱ_E is a negative when entering and a positive when exiting.

As shown if Fig. 2, the *magnetic ring* ϱ_M is North when exiting ($\equiv$ positive) and South ($\equiv$ negative) when entering. The magnetic field lines collapse to form a magnetic circle. The circle formation is not completed in Fig. 2. Note, though the magnetic circle is show as clockwise, both clockwise and anticlockwise orientations of the magnetic circle are possible.

Regarding stability, Fig. 3 shows how a *magnetic ring* ϱ_M, would be oriented within the electric field of a charged particle. The full magnetic circle is not shown in this cross section. The latent velocities caused by $E_Y \times B_Z$ are greater when θ is larger and lesser when θ is smaller.

Therefore, by *Ni Fields* [1], [4] & [6], there is an inward acceleration that keeps the rings in the central position and, this configuration is stable.

The electric ring would be similarly stable if its field was the opposite of the charged particle, then *unlike* directions of field lines would attract the *electric ring* ϱ_E and keep this ring in the central position. However, this would not be a sufficient condition as the electron has both positive and negative spins. As will be shown in section 7 & 8, energy is not a consideration, and stability is derived from the Ω structure, akin to the electron shell.

The proposed *probabilistic ring field* ϱ_ψ, consists of n integer multiples of space waves (2) & (9). One reason for a stable ϱ_ψ ring position would be it resides at a node in the *Probabilistic Wave Function* Ψ_P [14], and from (9) its ring radius $\varrho_{R,\psi}$ is,

$$\varrho_{R,\psi} = \left(\frac{1}{2} + n\right)\frac{\pi}{\lambda} \tag{50}$$

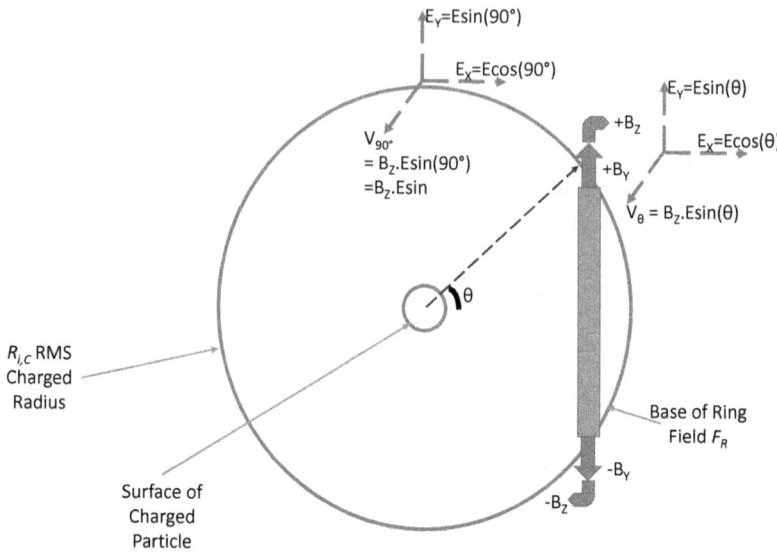

Fig. 3, The Magnetic Ring Field within an Electrically Charged Particle

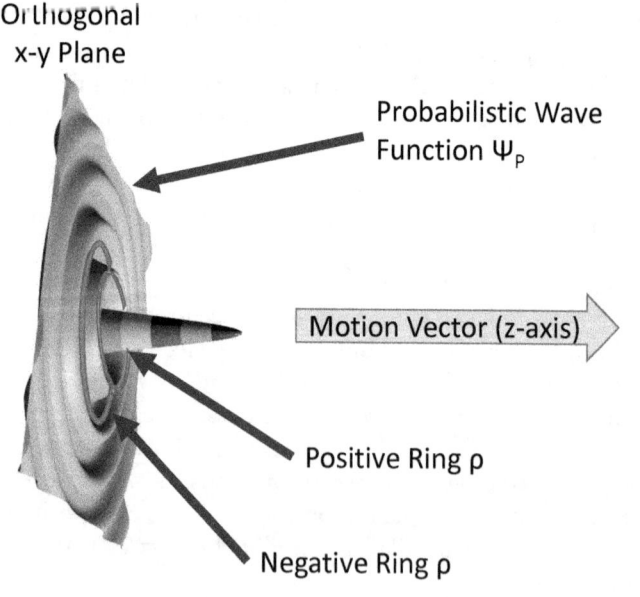

Fig. 4, Particle Structure.

Solomon [14] showed that the *Probabilistic Wave Function* Ψ_P is disc like, Fig. 4, in the *x-y* plane that is orthogonal (*z*-axis) to the motion of the particle. Therefore, one could assume that all *ring fields* ρ_F lie in this *Probabilistic Wave Function* Ψ_P *x-y* plane[vi].

Ring fields ρ_F is one mechanism as to how particles exhibit quantized physical properties. Adding a specific type of *ring field* ρ_F to a particle's structure increases that quantized amount by *1*.

However, to implement Pauli's Exclusion Principle[vii] with respect to electron spin, would require additional considerations. First that the *electric* ρ_E & *magnetic* ρ_M *ring fields* would sit at a crest or trough of the *Probabilistic Wave Function* Ψ_P. And from (9) the *crest ring radius* $\rho_{R,C}$ and *trough ring radius* $\rho_{R,T}$ are,

$$\rho_{R,C} = \left(\frac{1}{4} + n\right)\frac{\pi}{\lambda} \tag{51}$$

$$\rho_{R,T} = \left(\frac{3}{4} + n\right)\frac{\pi}{\lambda} \tag{52}$$

Where *n* = number of periods from *0* to some large value.

Using the nomenclature of building blocks, for the *electric ring field* ρ_E, the *field attenuation* $\rho_{A,E}$ at crest (53) and trough (54) and *field derivative* $\rho_{D,E}$ (55) are,

$$\rho_{A,EC} = sin\left(\left(\frac{1}{4} + n\right)\frac{\pi}{\lambda}\right) = 1 \tag{53}$$

$$\rho_{A,ET} = sin\left(\left(\frac{3}{4} + n\right)\frac{\pi}{\lambda}\right) = -1 \tag{54}$$

$$\rho_{D,E} = K_E \tag{55}$$

Where *electric ring constant* K_E = some constant, and by (6),

$$\rho_{B,E} = \pm K_E \tag{56}$$

This approach to reverse engineering a new Standard Model, the *Component Standard Model (CSM)*, will have some components similar to the Standard Model, and others different. Similar because it needs to fit the empirical data and different because its structure is different from the Standard Model.

Therefore, unlike the Standard Model's intrinsic spin *s* and fine structure s_z, the *CSM* only has *electric ring fields*, but to maintain some semblance, each *electric ring field* would be *equivalent* to quantum mechanics' ½ spin.

Table 1(a), Ring Field Analysis of Neutron Decay with +½ spin Electron

Particles	Original	→	Decay Products		Anti-neutrino *	Unknown Particle(s)
	Neutron	→	Proton	Electron		
Standard Model	n	→	p	e⁻		
Spin	½		½	½	½	-1
Electric Ring Fields						
Positive	1	=	1	1	1	
Negative		=				-2
Sub Total	1		1	1	1	2

Table 1(b), Ring Field Analysis of Neutron Decay with -½ spin Electron

Particles	Original	→	Decay Products		Anti-neutrino *	Unknown Particle(s)
	Neutron	→	Proton	Electron		
Standard	n	→	p	e⁻		
Spin	½		½	-½	½	0
Electric Ring Fields						
Positive	1	=	1		1	
Negative		=		-1		
Sub Total	1		1	1	1	0

Table 1(c), Ring Field Analysis of Neutron Decay with +1e Electron Magnetic Moment

Particles	Original	→	Decay Products		Anti-neutrino *ᶦ ₓ	Unknown Particle(s)
	Neutron	→	Proton	Electron		
Standard	n	→	p	e⁻	⁻νₑ	
Magnetic	1		≈-3/2n	+1e	+1⁻νₑ	1½
Magnetic Ring Fields						
North	2	=		1	1	3
South		=	-3			
Sub Total	2		3	1	1	3
Net Sign	+		-	+	+	+

Therefore, two electrons with *like* $\varrho_{B,E}$ (56) will repel and therefore cannot exist in the same state, while *unlike* will attract and do exists together in the same state.

Similarly, using the nomenclature of building blocks, for the *magnetic ring field* ϱ_M, the *field attenuation* $\varrho_{A,M}$ at crest (57) and trough (58) and *field derivative* $\varrho_{D,M}$ (59) are,

$$\rho_{A,MC} = sin\left(\left(\frac{1}{4}+n\right)\frac{\pi}{\lambda}\right) = 1 \tag{57}$$

$$\rho_{A,MT} = sin\left(\left(\frac{3}{4}+n\right)\frac{\pi}{\lambda}\right) = -1 \tag{58}$$

$$\rho_{D,M} = K_M \tag{59}$$

Where *magnetic ring constant* K_M = some constant, and by (6),

$$\rho_{B,M} = \pm K_M \tag{60}$$

Again this approach to reverse engineering a new Standard Model, the *Component Standard Model (CSM)*, would require a similarity with *electric ring fields* thus *magnetic ring fields* would be *equivalent* to quantum mechanics' ½ magnetic moments, but without including the particle's mass.

Setting each *electric ring field* $F_{B,E}$ to the same value as in quantum mechanics $s_{\vec{z}}$,

$$\rho_{B,E} = \rho_{A,E}\rho_{D,E} = (\pm 1)\left(\frac{\hbar}{2}\right) \tag{61}$$

Where *electric ring constant* (55) $K_E = \hbar/2$. To determine the values for the magnetic ring fields, the Stern–Gerlach experiment can be interpreted as

Table 2, Quark Ring Field Analysis of Neutron Decay

Particles	Original Up	Original Down	→	Decay Products Up	Decay Products Down	Unknown Particle(s)
Standard Model	u	2d	→	2u	d	
Spin	½	½		½	½	-1
Quark Model Electric Ring Fields						
Positive	1	2	=	2	1	
Negative			=			-2
Sub Total	1	2		2	1	-2
Magnetic Moment	1			0	-1½	1½
Magnetic Ring Fields						
North		2	=			3
South			=		-3	
Sub Total #Rings	0	2		0	3	3
Net Sign	+	+		+	-	+

follows, given the relationship[viii] [17] spin s_z and spin magnetic moment μ_{spin},

$$\mu_{spin} = \frac{e}{m_e} s_z \qquad (62)$$

or

$$\mu_{spin} = \frac{e}{m_e}\left(R_{A,M} R_{D,M}\right) = \frac{e}{m_e}(\pm 1)\left(\frac{\hbar}{2}\right) \qquad (63)$$

Where *magnetic ring constant* (59) $K_M = \hbar/2$. That is, the magnetic moments are quantized but the charge to mass ratios obfuscates this.

6.5 Component Standard Model

The *Component Standard Model (CSM)* is a different type of particle model from the Standard Model of quantum mechanics and should allow for a similar set of particles. To get some idea of how *CSM* works let use neutron decay as an example. See Table 1(a).

The Standard Model's spins are given in the 5th row. Since each electric ring is equivalent to ½ spin, the net of the positive and negative ring must be equivalent to the particle's spin in that column.

The fourth particle design note, a ring conservation law, requires that the sum of the rings before decay or reaction should equal the sum of the rings after decay or reaction. The fifth particle design note, is to have the minimum number of ring fields.

Therefore, the decay products have 3 ½ spins, or 3 positive rings, but for the neutron to have a ½ spin there needs to 2 negative rings (*3-2=1* or ½ spin). Since the decay products cannot accommodate the 2 negative rings, it must fall under a new column for an unknown particle. Given that 2 negative rings are equivalent to -1 spin, this would suggest that - given all other factors the same - the unknown particle(s) is like a photon as it will be massless.

What if the electron spin was -½? This is a possibility with *electric ring fields*.

Table 1(b) shows that if the decay electron had a -½ spin then, if there are any new unknown particle(s), its spin would be equivalent to 0.

Assuming that without including a particle's mass, magnetic moments are quantizable. Setting the neutron's magnetic moment equivalent to *1*, or *2 magnetic rings fields*, Table 1(c) provides the determination of these *ring fields*. The proton has a magnetic moment that is approximately *-3/2x* that of the neutron, therefore it has 3 rings if the neutron has 2. The negative sign requires 3 South, with the neutron having 2 North. By conservation of rings, this requires 3 Norths from an unknown particle(s).

Breaking down the decay into constituent quarks gives more information on quarks. Table 2 shows (excludes the electron and its electron antineutrino) that to get a minimum number of rings the magnetic moment of the up quark does not have magnetic ring fields or magnetic moments, while a down quark goes from 1 North ring to 3 South rings. The result is that the unknown particle(s) has the same magnetic moment of $1\frac{1}{2}$, and a strange particle.

If correct, it shows that quarks are stranger than thought, and the reason quarks are believed to be *confined* is that researchers are looking for quark particles with very different magnetic moments.

Though there is much work to be done on the *Component Standard Model*, its basic building blocks are *proto-fields* consisting of *null direction lines, ring, discs, spheres, kenoses*, electric and magnetic properties.

6.6 Hierarchy of Structure

Consolidating the previous discussion leads one to recognize a hierarchy of structure for matter.

At the first or lowest level, matter consists of *proto-fields, null direction lines ι, rings ρ, discs υ,* and *spheres ξ.* Based on their relationship to the *space wave* χ_P these *proto-fields* can take on positive or negative and north or south orientations for electric and magnetic properties, respectively.

At the second level are elementary particles, that consist of [1] motion vectors, and orthogonal discs of the particle's *Probabilistic Wave Function* Ψ_P (which is itself composed of the discs *space wave* χ_P and *envelope function* φ_P) and *electric ρ_E* & *magnetic ρ_M ring fields* and *electric $\xi_{P,E}$* & *magnetic $\xi_{P,M}$ spherical fields.* Assuming that the basic unit of charge is $\frac{1}{3}$ e, the electron would have 3 equal *electric proto-field spheres* $\xi_{P,E}$ with a radius of $\frac{3}{4}\lambda$ to give it a negative charge of $-e$. Similarly, a positron 3 *electric proto-field spheres* $\xi_{P,E}$ would have a radius of $\frac{1}{4}\lambda$ to give it a positive charge of $+e$.

The third level in this hierarchy are nucleon-like particles. Protons and neutrons are examples consisting of 'orbiting' quarks [2]. Even though charged, these quarks do not exhibit synchrotron radiation as their motion vectors are aligned with the centripetal acceleration, like open umbrellas with the handle pointing to the center of the nucleons. If particles have the magnetic monopole charge, it is not exhibited because its radius $\frac{1}{2}\lambda$ is at the node of the *space wave* χ_P.

Table 3, Electron-Positron Collision Analysis

Particles	Original		→	Decay Products		
	Electron e^-	Positron e^+	→	Photon γ^+	Anti-photon γ^-	Unknown Particle(s)
Standard Model						
Spin	½	-½		1	-1	
Electric Ring Fields						
Positive	1		=	2		
Negative		-1	=		-2	
Sub Total	1	1		2	2	
Mag. Mom.	≡½	≡-½		0?	0?	
Magnetic Ring Fields						
North		1	=	1	1	
South	-1		=	-1	-1	
Sub Total #Rings	1	1		2	2	
Net Sign	-	+				

The fourth level in this hierarchy are the nucleuses, consisting of combinations of nucleons and elementary particles.

The fifth level is the electron shell consisting of orbiting electrons [2]. Like quarks in nucleons, these electrons do not exhibit synchrotron radiation as their motion vectors, like open umbrellas, are aligned with the center of the nucleus.

From this hierarchy antimatter is defined as a wave function Ψ_{P^-} that completely cancels out the wave function Ψ_{P^+} of matter, or from (1),

$$\psi_{\bar{P}} = \varphi_P \chi_{\bar{P}} = \left(1/k_\psi x\right) sin\left(2\pi\frac{x}{\lambda} - \frac{\pi}{2}\right) \tag{64}$$

Or

$$\psi_{\bar{P}} = \varphi_P \chi_{\bar{P}} = \left(1/k_\psi x\right) cos\left(k_\psi x\right) \tag{65}$$

That is, as the wave function has shifted left by $\pi/2$ to become a cosine function, all particle properties that are related to the wave function are reversed. These include its charge, spin and magnetic moments. However, photons have both ±1 spins and therefore both photons and anti-photons are indistinguishable, but the altered wave functions should be evident in interference patterns.

Why then is there no such thing as anti-energy? By (1) the wave function is non integrable and therefore, cannot be the carrier of the particle's energy. By (34), (35), (36) and (37) the electric and magnetic field vectors rotate between *spacetime kenos* α and *subspace kenos* β. Their frequencies are determined by *Probabilistic Wave Function* Ψ$_P$, therefore, the rotation, not the vectors, of electromagnetic vectors is the energy of the photon.

However, per (22) and (28) the *electric* ι$_E$ and *magnetic* ι$_M$ *null direction line proto-fields* lie in the wave function disc, and are determined by the wave function. When the wave function is shifted from sine to cosine these *null direction line proto-fields* taken on opposite signs. However, from the perspective of the transverse wave and therefore the sense of the rotation between *spacetime kenos* α and *subspace kenos* β, these are not distinguishable. Therefore, the electromagnetic energy is the same for both the photon and the anti-photon. Table 3, details a possible ring field analysis of the low velocity electron-positron collision.

The inference from Tables 1, 2 & 3 is that since rings exists within the wave function which does not carry energy, energy is not required to change the nature of the rings, but there may be other factors that must hold.

6.7 A Proposal for Mass

From the perspective of *space waves*, Solomon [14] proposed how the mass equivalent energy is structured to determine the de Broglie wavelengths. The total *mass equivalent energy* E$_{m,T}$ of a particle with mass *m* traveling at a velocity *v*, in terms of its rest mass E$_{m,0}$ or rest mass term and its *kinetic energy mass* E$_{m,K}$ or energy term as,

$$E_{m,T} = E_{m,0} + E_{m,K} \tag{66}$$

$$E_{m,T} = m_0 c^2 + m_v v^2 \tag{67}$$

And therefore, applying relativistic corrections to (65), the total mass wavelength λ$_T$, the rest mass wavelength λ$_0$ and the kinetic energy wavelength λ$_K$ are given by,

$$\frac{h}{\lambda_T c} = \frac{h}{\lambda_0 c} + \frac{h}{\lambda_K v} \tag{68}$$

And

$$\lambda_0 = \frac{h}{m_0 c} \tag{69}$$

$$\lambda_K = \frac{h}{m_v v} = \frac{h}{m_0 v} \sqrt{1 - \frac{v^2}{c^2}} \tag{70}$$

123

Super Physics for Super Technologies

Table 4, Particle Ω Structures

Electron Ω Structure

| Location | | | | | | | | |
Component	E	M	Ψ	χ	φ	rest mass	kinetic mass
line ι							
ring ρ	±1	+1				+1	+1
disc υ			+1	+1	+1		
sphere ξ	-3						

Positron Ω Structure

| Location | | | | | | | | |
Component	E	M	Ψ	χ	φ	rest mass	kinetic mass
line ι							
ring ρ	±1	-1				-1	-1
disc υ			-1	-1	+1		
sphere ξ	+3						

Photon Ω Structure

| Location | | | | | | | | |
Component	E	M	Ψ	χ	φ	rest mass	kinetic mass
line ι	+1	+1					
ring ρ	+2	+2				+1	+1
disc υ			+1	|1	|1		
sphere ξ							

AntiPhoton Ω Structure

| Location | | | | | | | | |
Component	E	M	Ψ	χ	φ	rest mass	kinetic mass
line ι	-1	-1					
ring ρ	-2	-2				-1	-1
disc υ			-1	-1	+1		
sphere ξ							

Where (68) is the energy term that causes the *projected wave function* of interference patterns. That is, the particle's *true wave function* wavelength per (67) remains stable at any velocity while its *projected wave function* wavelength per (68) approaches zero as its velocity approaches the velocity of light.

By (68) there are two parts to mass, rest and kinetic, and both are exhibited as *space waves* $\chi \rho$. However, the space wave is not a carrier of energy, therefore, both mass and kinetic energy should be a form of electromagnetic energy as energy is the rotation of the electric and magnetic vectors.

This paper proposes that the rest mass is an *electromagnetic ring field* within the wave function. Its wavelength is given by (69). The kinetic energy mass is stored as a second *electromagnetic ring field* whose wavelength is determined by (70).

The *electromagnetic ring field* consists of the electric and magnetic vectors rotating between spacetime and subspace in a circular path within the wave function.

Therefore, the particle remains an orthogonal disc to its motion vector while its spherical properties are derived from its *spherical proto-fields*.

Thus, like electric charge, the *rings fields* placed at a radii of $\frac{3}{4}\lambda$ would be a negative and at $\frac{1}{4}\lambda$ would be a positive. However, this would be evidenced as a phase shifted electric and magnetic vector rotations. These phase shifted rotations are indistinguishable. That is, negative mass cannot exists, and all mass is positive electromagnetic.

Solomon [1] proposed a definition for localization that particles emerge from the *subspace kenos*. One can propose a more complete definition. When a particles exits the *spacetime kenos a* and submerges into the *subspace kenos β* all its *null direction lines*, *rings* and *spheres* shift to the nodes of the *space wave χ$_P$*, and are no longer discernible. All that remains observable is the *Probabilistic Wave Function* Ψ_P. On localization all its *null direction lines*, *rings* and *spheres* shift back to their original positions in the *space wave χ$_P$*, and emerge from *subspace kenos β* back into the *spacetime kenos a*.

This process therefore requires a particle to have structure Ω information, equivalent to the electron shell. This structure allows the particle's components to occupy specific states in the *spacetime kenos a* out of three, -1, 0 & +1, and the 0 state in the *subspace kenos β* as determined by the *space wave χ$_P$* trough, node & crest respectively. Therefore, the particle Ω structure can be defined as a set of particle components. Table 4, illustrates this Ω structure for an electron, positron, photon and anti-photon.

Note, Solomon [1] proposed two types of mass rest mass and kinetic mass that are formed by two different *electromagnetic rings*. The blue shaded boxes indicate no change in final property if location is changed between +1 and -1. For example, the electromagnetic mass ring ϱ_{em} does not change sign if it switches its space wave location between crest +1 and trough -1. Yes, there probably is an equal amount of anti-energy and anti-mass but the Universe cannot tell the difference. The other assumption is that the basic unit of charge is $\frac{1}{3}\ e$.

6.8 Conclusion

By proposing a component approach to constructing particles this paper has developed a building block approach to particle design. Thereby showing why there is no such thing as anti-energy or anti-mass. The proposed Component Standard Model allows for new types of particles to be added by just adding new properties.

References:

[1] Solomon, B.T. "Replacing Schrödinger", in *Super Physics for Super Technologies*, March 2015.

[2] Solomon, B.T. "A Non Standard Model Nucleon/Nuclei Structure", in *Super Physics for Super Technologies*, March 2015.

[3] Solomon, B. T., "The Variable Isotopic Gravitational Constant", in *Super Physics for Super Technologies*, March 2015.

[4] Solomon, B.T., "A Universal Approach to Forces", in *Super Physics for Super Technologies*, March 2015.

[5] Solomon, B. T., "Empirical Evidence Suggest A Need For A Different Gravitational Theory," in the proceedings of the *100 Year Starship Study Public Symposium (100YSS,)*, 2013.

[6] Solomon B.T., "New Evidence, Conditions, Instruments & Experiments for Gravitational Theories", Journal of Modern Physics, Special Issue on Gravitation, Astrophysics and Cosmology, Vol. 8A, 2013, August 2013.

[7] Solomon, B. T., "Empirical Evidence Suggest A Need For A Different Gravitational Theory," *American Physical Society (APS) April Conference, Denver*, 2013

[8] Solomon B.T., *An Introduction to Gravity Modification: A guide to using Laithwaite's and Podkletnov's experiments and the physics of forces for empirical results.* Universal Publishers, Boca Raton, 2nd Edition, May 2012.

[9] Solomon, B. T., "Non-Gaussian Radiation Shielding," *100 Year Starship Study Public Symposium (100YSS,)*, 2011.

[10] Solomon, B.T., "Gravitational Acceleration Without Mass And Noninertia Fields", Physics Essays, Vol. 24, 327, 2011. [Phys. Essays **24**, 327 (2011)]

[11] Solomon, B. T., "Reverse Engineering Podkletnov's Experiments," in the proceedings of the *Space, Propulsion & Energy Sciences International Forum (SPESIF-11)*, Edited by Glen A Robertson, Physics Procedia, Elsevier Science.

[12] Solomon, B. T., "Non-Gaussian Photon Probability Distributions," in the proceedings of the *Space, Propulsion & Energy Sciences International Forum (SPESIF-10)*, Edited by Glen A Robertson, AIP Conference Proceedings **1208**, Melville, New York, (2010).

[13] Solomon, B. T., "An Approach to Gravity Modification as a Propulsion Technology," in the proceedings of the *Space, Propulsion & Energy Sciences International Forum (SPESIF-09)*, Edited by Glen A Robertson, AIP Conference Proceedings **1103**, Melville, New York, (2009).

[14] Solomon, B. T., "An Epiphany On Gravity", Journal of Theoretics, Vol. 3-6, 2001.

[15] Feynman, R., *Feynman Lectures on Physics*, Chapter 28-4.

[16] Nemiroff, R., "Bounds on Spectral Dispersion from Fermi-Detected Gamma Ray Bursts," Physical Review Letters, Vol. 108, No. 23, 2012.

[17] John Taylor, Modern Physics for Scientists and Engineers (2nd Edition), 2003.

[18] Daraktchieva, Z., The New Limit for the Magnetic Moment of the electron antineutrino from MUNU Experiments, http://arxiv.org/ftp/hep-ex/papers/0305/0305057.pdf

End Notes:

i.   Does not include non-peer reviewed conference presentations between 2001 and 2009.

ii.   Sure, the Standard Model approach to monopoles http://www.theory.caltech.edu/~preskill/pubs/preskill-1984-monopoles.pdf has a slightly different equation. It is based on a Schrödinger wave function quantum mechanical model that Solomon [1] has shown does not fit the empirical data of the interference pattern exactly. 'Exactly' is a question of tolerance, if the required R^2 is 99.9999% then it is not exact. Further, monopoles are not verifiable (as opposed to falsifiable) because it requires a particle accelerator so massively powerful that it is unlikely that any organization could build. Therefore, the proposed monopole equation, mimics that of the electric charge, with further details being worked out in future papers.

iii. http://hyperphysics.phy-astr.gsu.edu/hbase/electric/engfie.html

iv.   http://physics.nist.gov/cgi-bin/cuu/Value?mupsmunn

v.   Acceleration is present where there is a spatial gradient of real or latent velocities.

This acceleration is along the gradient of greater velocities, and not dependent upon the direction of these velocities.

vi.   This suggests that it is the spherical electrical fields that cause particles to appear spherical.

vii.   http://hyperphysics.phy-astr.gsu.edu/hbase/pauli.html

viii.   John Taylor, Modern Physics for Scientists and Engineers (2nd Edition), equations 9.26 & 9.27

ix.   The electron antineutrino has a magnetic moment [18] http://arxiv.org/ftp/hep-ex/papers/0305/0305057.pdf

x.   http://physics.nist.gov/cgi-bin/cuu/Value?mupsmunn

7. Spectrum Independence

Abstract:- Based on empirical data, this paper proposes and defines shielding, wrapping or the ability to get around an obstruction, and invisibility are the same phenomenon. Clarifying that cloaking, resolution and transmission are variations of wrapping.

One of the components of the probabilistic (not the Schrödinger) wave function, is the non-Gaussian probabilistic *envelope function* φ_P [2]. It is this property that enables the unification of shielding, wrapping and invisibility into a single phenomenon, and therefore, results in the spectrum independent photon analytics.

The shielding hypothesis was tested against NASA's published microwave empirical data with excellent fit [18]. The published nanowire energy distribution data was used as a benchmark for the transmission hypothesis with excellent agreement.

Further, this approach proposed a method to determining propagation distance in all coaxial type structures from nanowires to coaxial cables. The net result is a consistent and uniform set of formulae that determine photon behavior across the electromagnetic spectrum.

This approach allows macro photon phenomena to be constructed from particle properties, and a first step towards directly unifying quantum level properties with macro level phenomena.

7.1 Introduction

This paper reports the 16th paper in the 16-year ([1] to [15][i]) investigation into the feasibility of gravity modification, and a *New Standard Model*.

Primarily, that the spatial gradients of fields is a universal mechanism by which forces are transmitted [5], the Schrödinger wave function [2] has a replacement (1),

$$\psi_P = \varphi_P \chi_P = \left(1/k_\psi x\right)sin\left(k_\psi x\right) \tag{1}$$

Where the *space wave* χ_P is,

$$\chi_P = sin\left(k_\psi x\right) \tag{2}$$

And the *envelope function* φ_P is

$$\varphi_P = \left(1/k_\psi x\right) \tag{3}$$

Poincaré stresses [16] do not exist [5], and quantum vacuum does not exists [17], the new *Component Standard Model* [1], the gravitational constant G is a variable [4] and there exists a massless formula for gravitational acceleration g (5),

$$k_\psi = \left(2\pi/\lambda\right) \tag{4}$$

$$g = \tau c^2 \tag{5}$$

where τ is the spatial gradient, the change in time dilation divided by the change in that distance.

These findings should lead to the replacement of the Standard Model with something substantially simpler. Occam's Razor, that the simpler solution is most likely the correct solution, affirms this investigation.

7.2 Geometric Relationships

This paper proposes photon physics is spectrum independent as it is a function of the geometric structure in the photon's environment. Solomon [8] & [1] had proposed that particle probabilities are non-Gaussian, and that such an approach would lead to nanowire energy profiles that were similar to quantum mechanical models [18]. Building on these, this paper shows that shielding, wrapping[ii] and invisibility (SWI) are different manifestations of the same geometric phenomenon. Transmission is a variant of invisibility, with nanowire, microwave and radio wave transmission handle by the same formula. Therefore, the term Photon SWI or PSWI.

Solomon [1] & [2] proposed that the electromagnetic wave is a flat disc originating from the *Probabilistic Wave Function* Ψ_P (1), also the solution ψ_P, and therefore consistent with Lorentz-FitzGerald transformations (LFT) and Special Theory of Relativity (STR). To be consistent with conservation of mass energy [1], the electric and magnetic vectors rotate between the x, y, z, t, *spacetime kenos* (region) a and the x, y, z, subspace kenos β that are 90° out of phase.

To distinguish between the photon and the electromagnetic wave, in addition to the transverse electromagnetic wave, the photon also consists of three other building blocks [1] the probabilistic R_ψ, electric R_E, and magnetic R_M *ring fields* R_F but these are not relevant to this paper.

The transverse electromagnetic (TEM) wave can be described using the Building Block [1] nomenclature. The TEM as a building block F_B (6) consists of two field components, the *field attenuation* F_A and *field derivative* F_D.

$$F_B = F_A F_D \tag{6}$$

The *field derivative* $F_{D,EW}$ [1] of an oscillating electric vector of maximum strength E_A, and wavelength λ, propagating a distance z, is given by,

$$F_{D,EW} = E_A \tag{7}$$

And its *field attenuation* $F_{A,EW}$ [1] is,

$$F_{A,EW} = sin(k_{B,\psi} z) \tag{8}$$

$$k_{B,\psi} = (2\pi/\lambda) \tag{9}$$

Where $k_{B,\psi}$ is the *probabilistic field attenuation constant*. Therefore, by (6), the oscillating electric wave $F_{B,EW}$ is,

$$F_{B,EW} = E_A sin(k_{B,\psi} z) \tag{10}$$

Similarly, oscillating magnetic wave of maximum strength B_A, is described by $F_{B,MW}$ and defined by its *field derivative* $F_{D,MW}$ and *field attenuation* $F_{A,MW}$, thus,

$$F_{D,MW} = B_A \tag{11}$$

$$F_{A,MW} = sin(k_{B,\psi} z) \tag{12}$$

$$F_{B,MW} = B_A sin(k_{B,\psi} z) \tag{13}$$

Note, TEM consists of electric (10) and magnetic (13) spatial oscillations along the z-axis, that are derived from the *Probabilistic Wave Function* Ψ_P. Since TEM oscillates between *spacetime kenos* a and the *subspace kenos* β, it is attenuated by the probabilistic *envelope function* φ_P (3).

This geometric structure enables a geometric definition for shielding Θ_S, wrapping Θ_W and invisibility Θ_I. These definitions are set up to evidence a *monotonic* relationship, i.e. a larger value for a greater effect.

Shielding effectiveness or just *shielding* Θ_S is the ability of a shield of radius R to prevent a photon pass through an aperture of radius r in that shield.

Defined in decibels as the ratio of the function of the total photon probability $f(P_{\leq R})$ across the shield and aperture, to the function of the probability $f(P_{\leq r})$ that the photon will pass through the aperture,

$$\Theta_S = 10log\left(\frac{f(P_{\leq R})}{f(P_{\leq r})}\right)$$

(14)

Wrapping effectiveness or just *wrapping* Θ_W is the ability of a photon to get around an obstruction of radius r in a much larger aperture of radius R. Defined[iii] in decibels as the ratio of the function of the probability $f(P_{>r})$ that the photon will pass around the obstruction to the function of the total photon probability $f(P_{\leq R})$ across the aperture and obstruction,

$$\Theta_W = 10log\left(\frac{f(P_{>r})}{f(P_{\leq R})}\right)$$

(15)

Note, the term *wrapping* is used to distinguish this phenomenon from cloaking. Cloaking requires an additional requirement, that TEM propagate relatively undisturbed, around the obstruction. Like *cloaking, transmission* as used in coaxial cables and nanowires, is variation of *wrapping*.

Resolution Θ_R is the opposite of *wrapping*. *Resolution* Θ_R is the smallest object size that obstructs/reflects the photon. *Resolution effectiveness* or just *resolution* Θ_W is the ability of a photon to be obstructed by and object of radius r in a much larger aperture of radius R. Defined in decibels as the ratio of the function of the probability $f(P_{\leq r})$ that the photon will be ob-

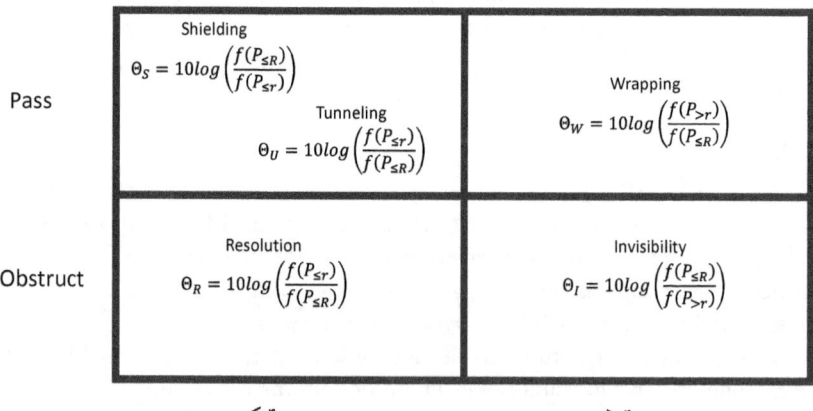

Fig. 1, *The Relationships between Shielding, Wrapping, Resolution and Invisibility*

structed to the function of the total photon probability $f(P_{\leq R})$ across the aperture and obstruction, and is defined as,

$$\Theta_R = 10 log \left(\frac{f(P_{\leq r})}{f(P_{\leq R})}\right) \tag{16}$$

Invisibility effectiveness or just *invisibility* Θ_I is the ability of a photon to pass through an aperture of radius r in a shield of radius R. Defined in decibels as the ratio of the function of the total photon probability $f(P_{\leq R})$ across the shield and aperture to the function of the probability $f(P_{>r})$ that the photon will not pass through the aperture,

$$\Theta_I = 10 log \left(\frac{f(P_{\leq R})}{f(P_{>r})}\right) \tag{17}$$

Quantum tunneling Θ_U can defined as the ratio of the function of the function of the probability $f(P_{\leq r})$ that the photon will pass through the apertures between the atoms and molecules, to the total probability $f(P_{\leq R})$ that the photon will localize on the atoms, molecules and not localize at the aperture. Therefore, quantum tunneling is the opposite of resolution.

Summarizing these four phenomena in Fig. 1, it is clear that these are complimentary phenomena. That is, these are four different variations of the same phenomenon.

Equations (14) to (17) benchmark both the physics and technology with respect to random events. For shielding, wrapping, resolution and invisibility to be greater than just probability alone, these need to exceed the value provided by (14), (15), (16) & (17), respectively.

In free space as $R \rightarrow \infty$, $P_{\leq R} = 1$ and therefore, these equations can be written as,

$$\Theta_{S,\infty} = 10 log \left(\frac{f(1)}{f(P_{\leq r})}\right) \tag{18}$$

$$\Theta_{W,\infty} = 10 log \left(\frac{f(P_{>r})}{f(1)}\right) \tag{19}$$

$$\Theta_{R,\infty} = 10 log \left(\frac{f(P_{\leq r})}{f(1)}\right) \tag{20}$$

$$\Theta_{I,\infty} = 10 log \left(\frac{f(1)}{f(P_{>r})}\right) \tag{21}$$

7.3 Testing This Hypothesis

Otoshi [18] provides an experimental set up using microwaves transmissions through a waveguide (length l, height h & width w) blocked by a perforated flat plate to test shielding. A numerical model was built to compare Otoshi's empirical results, quantum mechanical theoretical, and Photon SWI (PSWI) hypothesis presented in this paper.

The electric E_A and magnetic B_A field strengths within the TEM is determined by electric η_E and magnetic η_M field energy densities[iv] given by

$$T\eta_E = \alpha\eta_E + \beta\eta_E = \left(\frac{\varepsilon E_A^2}{2}\right) \tag{22}$$

$$T\eta_M = \alpha\eta_M + \beta\eta_M = \left(\frac{B_A^2}{2\mu}\right) \tag{23}$$

Total [1] electrical energy density $T\eta_E$ and total magnetic energy density $T\eta_M$ are the sum of their respective components in the *spacetime kenos a* and *subspace kenos* β, such that these total energies are always a constant, or the photon E_P in the volume of space it occupies, its energy density $T\eta_P$,

$$T\eta_P = T\eta_E + T\eta_M = \left(\frac{\varepsilon E_A^2}{2}\right) + \left(\frac{B_A^2}{2\mu}\right) \tag{24}$$

Making use of the fact[v] that the ratio of the field amplitudes equals the velocity of light,

$$\frac{E_A}{B_A} = c \tag{25}$$

Gives,

$$T\eta_P = (1/\mu)B_A^2 = \varepsilon E_A^2 \tag{26}$$

Or

$$E_A = \sqrt{T\eta_P/\varepsilon} \tag{27}$$

$$B_A = \sqrt{T\eta_P\mu} \tag{28}$$

This paper proposes that, *without increasing the total energy*, there are only two mechanism for increasing the energy density in a confined space, these are probabilities and reflections.

Going back to first principles the *field attenuations* are the cross sectional probabilities $P_{(x,y,z)}$ at location (x, y, z). The attenuated electric $E_{A,(x,y,z)}$ and magnetic $B_{A,(x,y,z)}$ field strengths can be written as,

$$E_{A,(x,y,z)} = P_{(x,y,z)}E_A \tag{29}$$

$$B_{A,(x,y,z)} = P_{(x,y,z)}B \tag{30}$$

Therefore, the probabilistic energy densities transmitted through the perforated metal plate's apertures are,

$$_T\eta_T = \varepsilon\left(P_{(x,y,z)}E_A\right)^2 = \frac{1}{\mu}\left(P_{(x,y,z)}B_A\right)^2 \tag{31}$$

Governed by the *probabilistic envelope function* φ_P, the photon is less likely to pass through an aperture radially further from its axis of propagation then one that is closer.

Since the *probabilistic wave solution* ψ_P, has this radial drop off, the normalized wave function probability $P\psi_{,(x,y,z)}$ at any point (x, y, z) [2] would take the form,

$$P_{\psi,(x,y,z)} = \psi_{Pi,(x,y,z)}/F_{I,A} \tag{32}$$

Where the *field of interaction* $F_{I,A}$ at the cross sectional area of the perforated plate, limits the integration to this cross section and r is from 0 or at the axis of propagation, to the radial distance $L = h \ or \ w$ of the waveguide walls,

$$F_{I,A} = \int_0^L \pi\psi_P^2 dr \tag{33}$$

Or,

$$P_{\psi,(x,y,z)} = \psi_{Pi,(x,y,z)}\Big/\int_0^L \pi\psi_P^2 dr \tag{34}$$

But *probabilistic wave solution* ψ_P fluctuates with orthogonal radius. Therefore, this *probabilistic wave solution* ψ_P cannot attenuate this photon energy density as the sine term (1) has both positive and negative values, causing the energy density to acquire positive and negative values. That is (34) is not allowed. This confirms [2] the finding that *Probabilistic Wave Function* Ψ_P is non integrable into a mass-energy function. Considering that (34) is not allowed, this photon energy density, is only carried by the TEM in the photon.

The *probabilistic envelope function* φ_P on the other hand, does not have negative values, thus the normalized probability $P\varphi_{,(x,y,z)}$ at any point (x, y, z) [2] would take the form,

$$P_{\varphi,(x,y,z)} = \varphi_{Pi,(x,y,z)}/F_{I,A} \tag{35}$$

Where the *field of interaction* $F_{I,A}$ [2] at the cross sectional area of the perforated plate is,

$$F_{I,A} = \int_0^L \pi\varphi_P^2 dr_s \tag{36}$$

Where radial distance $L = h \ or \ w$ of the waveguide walls.

The radial drop off of this TEM energy over the cross section of the waveguide is governed by the normalized probabilistic *envelope function* φ_P (3) or $P\varphi_{,(x,y,z)}$ (35). Rewriting (31) from (35),

$$T\eta_T = \varepsilon\left(P_{\varphi,(x,y,z)}E_A\right)^2 = (1/\mu)\left(P_{\varphi,(x,y,z)}B_A\right)^2 \tag{37}$$

The numerical model of (37) confirms the need to include reflections as the estimated *effective shielding* Θ_S is larger ($59dB < \Theta_S < 73dB$) than the Otoshi data ($23dB < \Theta_S < 59dB$).

That is, since the microwave TEM energy is inside an enclosed metallic cavity it reflects back and forth. This raises the photon probabilities φ_P, by n number of times the electric field is reflected back *to the* perforated flat plate exit [22].

Therefore, by (29), (30) & (37) the transmitted microwave energy density $T\eta_T$ can be rewritten as,

$$T\eta_T = \varepsilon\left(nP_{\varphi,(x,y,z)}E_A\right)^2 = \frac{1}{\mu}\left(nP_{\varphi,(x,y,z)}B_A\right)^2 \tag{38}$$

By (14) shielding effectiveness Θ_S is the ratio of the photon energy E_v ($=h\nu$) to the transmitted energy E_T,

$$\Theta_S = 10log\left(\frac{E_v}{E_T}\right) \tag{39}$$

Since (39) is a ratio, the volume term in the energy density cancels, thus,

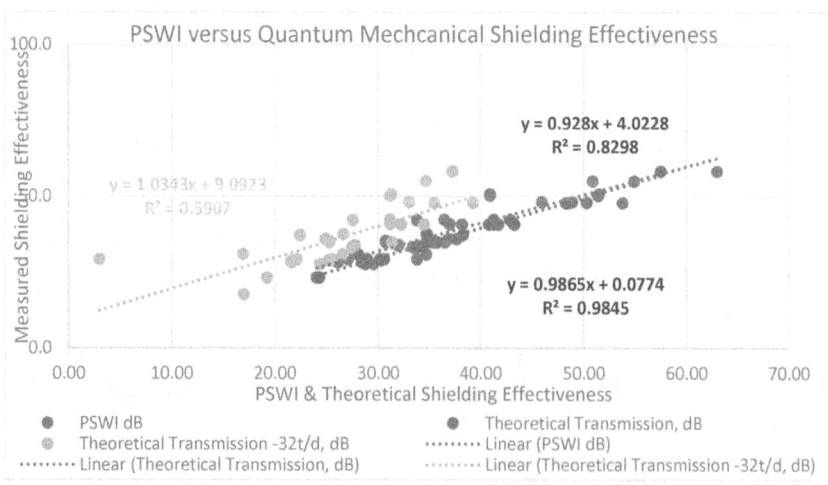

Fig. 2, Comparison of Quantum Mechanical versus PSWI Shielding Effectiveness

$$\Theta_S = 10log\left[\varepsilon E_A^2 / \varepsilon\left(nP_{\varphi,(x,y,z)}E_A\right)^2\right] \tag{40}$$

$$\Theta_S = 10log\left[1/\left(nP_{\varphi,(x,y,z)}\right)^2\right] \tag{41}$$

The numerical model was extended to include the number of reflections n per (41) and within the limits of Otoshi's data, a relationship was determined for *reflections n* in term of aperture *diameter d, wavelength λ,* and *porosity p* (defined as the ratio of the area of apertures to the cross sectional area of the perforate metal plate),

$$n = k_p p + k_d (d/\lambda) + k_n \tag{42}$$

Where the *porosity constant* k_p= *30.585327,* the *aperture ratio constant* k_d= *856.518003* and, the *reflection constant* k_n= *-28.059282.* The model shows, for the Otoshi data, the range of values n takes were,

$$1 < n < 82 \tag{43}$$

Which means that the non-transmitted photons were absorbed by the cavity walls. The R^2=*88.49%* for (42), is decent but not exciting.

With this PSWI approach the thickness of the metal plate is not statistically significant no matter what model relationship was tested.

The quantum mechanical [18] theoretical relationship Θ_{sq} for shielding effectiveness is,

$$\Theta_{SQ} = 10log\left[1 + \frac{1}{4}\left(\frac{3ab\lambda}{\pi d^3 cos\theta}\right)^2\right] + \frac{32t}{d} \tag{44}$$

Where a and b are spacings between apertures and θ is the angle of incidence to the metal plate.

Θ_{sq} relationship to the empirical data had an R^2= *98.44%,* see Fig 2., which is very good, but taking out the *32t/d* term causes the R^2 to drop significantly to *56.41%.* That is, though the *32t/d* term is essential to the quantum mechanical relationship, it loses its significance in the PSWI relationship. The PSWI inference is that photon arrivals to the metal perforated plate are independent of subsequent passage through the apertures, but this is not the case with the quantum mechanical interpretation

Returning to the earlier results without reflections, in free space the *shielding effectiveness* Θ_S would be determined by (41) with *n=1.* That is, confinement substantially increases transmission through the apertures.

Though further research is necessary to improve (42), the empirical data confirms that,

i) The PSWI approach to shielding and thus wrapping and invisibility are valid.

ii) The probabilistic function, *envelope function* φ_P (4) determines how both electric and magnetic energy densities reduce radially from the axis of photon propagation.

iii) That probabilities and reflections (41), are two mechanism for increasing the *energy density* in a confined space.

7.4 Wrapping & Invisibility

Using the mathematical development in the previous section, *wrapping effectiveness* Θ_W is defined as the ratio of the energy that continues to propagated E_P, around the obstruction to the photon energy E_v to,

$$\Theta_W = 10log\left(\frac{f(P_{>r})}{f(P_{\leq R})}\right) = 10log\left(\frac{E_P}{E_v}\right) \tag{45}$$

But the energy propagated E_P is total energy E_v less the obstructed energy E_O,

$$E_v = E_P + E_O \tag{46}$$

The obstructed energy density $\tau\eta_O$ is equivalent to the transmitted in shielding, or,

$$_1\eta_U - \varepsilon\left(\Gamma_{\varphi,(x,y,z)}E_A\right)^2 - (1/\mu)\left(P_{\varphi,(x,y,z)}B_A\right)^2 \tag{47}$$

In free space ($n=1$) this E_P/E_v ratio can be written in terms of the energy that is obstructed E_O and is equivalent to (37),

$$\frac{E_P}{E_v} = \frac{E_v - E_O}{E_v} = 1 - \left(P_{\varphi,(x,y,z)}\right)^2 \tag{48}$$

However, in a confined space (for example, a microwave wave guide with an obstruction in the middle) the propagated energy density cannot increase as it exits the confined space. That is, reflections only increase the obstructed energy. This obstructed photon energy density probabilities $P_{\varphi,(x,y,z)}$ is increased by the number of reflections n and (47) can be rewritten as,,

$$\Theta_W = 10log\left[1 - \left(nP_{\varphi,(x,y,z)}\right)^2\right] \tag{49}$$

That is internal reflections, within the obstruction, hampers *wrapping*.

Similarly, *invisibility effectiveness* Θ_I ratio is the inverse of *wrapping effectiveness* Θ_S ratio and is defined as, ratio of the obstructed energy E_O that does not propagate through the aperture to the total E_v,

$$\Theta_I = 10log\left(\frac{E_O}{E_v}\right) \tag{50}$$

The obstructed energy E_O is equivalent to the transmitted energy E_T in (39),

$$\frac{E_O}{E_v} = \frac{E_T}{E_v} = \left(nP_{\varphi,(x,y,z)}\right)^2 \tag{51}$$

Or,

$$\Theta_I = 10log\left[\left(nP_{\varphi,(x,y,z)}\right)^2\right] \tag{52}$$

The key to invisibility is to squeeze the *envelope function* φ_P (4) to effect a *virtual confinement* such that, rewriting (4),

$$\varphi_P = \left(k_c/k_\psi x\right) = (k_c\lambda)/(2\pi x) \tag{53}$$

Where the *confinement constant* k_c < 1. For invisibility to be a real technology, the process of making the photon pass through an object must be reversed on exiting the object. The Airy disc aperture diameter w_A provides clues. From [2], the projected *envelope function* φ is

$$\varphi = 1/u = 1/\left(\pi(w_A/\lambda)sin(\theta)\right) \tag{54}$$

or

$$\varphi = \frac{\lambda}{\pi w_A} \frac{1}{sin(\theta)} \tag{55}$$

Equation (55) shows increasing the Airy disc aperture diameter w_A reduces the *projected probabilistic envelope function* φ and spread of the interference patterns. Thus, reducing the wavelength λ could be an option. Applying a *virtual confinement* k_c is similar to increasing the aperture diameter w_A, would be better.

7.5 Subwavelength Confinement

Revisiting [8], Oulton *et al* [18] researched THz (λ=1,550nm) subwavelength confinement. Their experiments show that one can increase propagation distance while maintaining moderate confinement by tuning the geometric properties of the encased nanowire.

Subwavelength confinement, is photon propagation along a dielectric cylindrical GaAs nanowire of length l, radius r_d (or diameter d) embedded in SiO_2 at a distance h (from outer edge of nanowire) above a metallic region using hybrid waveguide. Since the λ/d ratios are large ($3{\leq}\lambda/d{\leq}15$)[vi], the optical photon in a nanowire is equivalent to a confined electromagnetic field in a radio antenna like a coaxial cable.

This paper proposes,

i) To exhibit a consistent set of behaviors across all phenomena, transmission by subwavelength confinement is a variation of wrapping with the additional requirement of (ii).

ii) That the photon has localized within the nanowire (diameter of 200 to 400 nm) and its *space wave* χ_P is trapped within the nanowire, it therefore, propagates along the nanowire. Per particle structure [1] & [2] one could propose that trapping occurs when the nanowire radius is greater than one *space wave* period λ (wavelength between 400 to 700 nm), if not the photon is not trapped and escapes. Thus its propagation depth is unpredictable and limited.

iii) However, its *envelope function* φ_P permeates the nanowire into the medium surrounding this nanowire. Like the microwave in the dielectric medium of the coaxial cable, the photon's electromagnetic wave propagates through the SiO_2 dielectric medium.

Building from first principles, the particle's wave function Ψ_P (2) [1] & [2] has zero physical thickness in the direction of propagation. Per (35) the *Field of Interaction* denominator term is the cross sectional area of the hybrid waveguide that is orthogonal to photon propagation. For rectangular waveguide with a height (from center of nanowire) above the nanowire of L and below of r_d+h, width on either sides of the nanowire of L, the total confined *envelope function* φ_P is given by,

$$F_{I,A} = \int_{-L}^{+L} \int_{d+h}^{L} \varphi_P \, dy \, dx \tag{56}$$

Therefore, the normalized probability $P_{NP,(x,y,z)}$ at (x, y, z) in a confined rectangular waveguide is,

$$P_{\varphi,(x,y,z)} = \varphi_{P,(x,y,z)}/F_{I,A} \tag{57}$$

$$P_{\varphi,(x,y,z)} = \varphi_{P,(x,y,z)}/\int_{-L}^{+L} \int_{rd+h}^{L} \varphi_P \, dy \, dx \tag{58}$$

The skin effect[vii] is "where alternating current avoids travel through the center of a solid conductor, limiting itself to conduction near the surface. This effectively limits the cross-sectional conductor area available to carry alternating electron flow, increasing the resistance of that conductor above what it would normally be for direct current".

In nanowires this effect is important as it forces electric field out into the dielectric medium. This is also true for coaxial cables. Like coaxial cables the nanowires would require a means for a return current through the metal plane.

The nanowire is treated as a cylindrical electrified surface (also known as the skin effect because this electric field is fluctuating like an AC electric field) [19] with field lines stretching out to the metal plane due to the transverse electromagnetic wave and therefore the electrical field strength decreases inversely with distance from the edge of the nanowire to the metal plane.

Using the building block nomenclature, for a charge per unit length q the electric field in the dielectric medium and outside the cylindrical nanowire $E_{A,d}$ (radius r_d, SiO$_2$ dielectric constants $\varkappa_d=2.25$) at a radius r is,

$$E_{A,d} = (1/(\varepsilon_d 2\pi r))q \tag{59}$$

The electric field $E_{A,n}$ inside the cylindrical nanowire (radius r_n, GaAs dielectric constants $\varkappa_n=12.25$) at a distance r is given by,

$$E_{A,n} = (r/\varepsilon_n 2\pi r_n^2)q \tag{60}$$

Keeping it simple, the electric field E_M in the lower metallic plane is negligible as its dielectric constant $\varkappa_M=129$.

$$E_{A,m} = 0 \tag{61}$$

From (31) the energy densities in the nanowire η_n, SiO$_2$ dielectric η_d and the metallic plane η_m are given by their respective electric field strengths,

$$\eta_n = \varepsilon_n \left(P_{\varphi,(x,y,z)}E_{A,n}\right)^2 \tag{62}$$

$$\eta_d = \varepsilon_d \left(P_{\varphi,(x,y,z)}E_{A,d}\right)^2 \tag{63}$$

$$\eta_m = \varepsilon_m \left(P_{\varphi,(x,y,z)}E_{A,m}\right)^2 = 0 \tag{64}$$

By wrapping (45) & (46) the propagated energy E_P, is the energy in the dielectric E_d. It is the difference between total energy E_v and energy inside the nanowire E_n, as $E_{A,m}$ (61) is zero

$$E_P = E_v - E_n \tag{65}$$

Therefore, energy as the sum of the energy densities across the cross sectional area (the length l of the nanowire cancels out) of the nanowire A_n, and dielectric medium A_d is formed by the radius r orthogonal to the nanowire,

$$E_n = \int_0^{r_d} \pi\varepsilon_n \left(P_{\varphi,(x,y,z)}E_{A,n}\right)^2 r^2 dr \tag{66}$$

$$E_d = \int_{r_n}^{L} \pi\varepsilon_d \left(P_{\varphi,(x,y,z)}E_{A,d}\right)^2 r^2 dr \tag{67}$$

And r cuts off at the metal plane r_n+h and at the edge of the dielectric medium L.

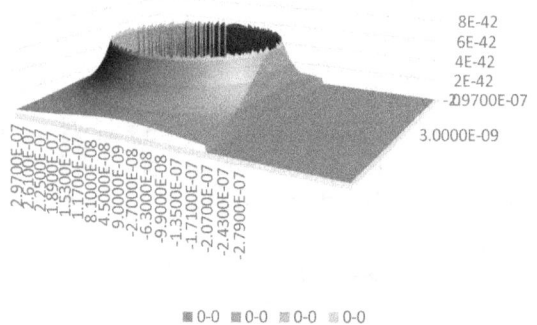

SiO$_2$ Electric Field Energy Density

Fig. 3, SiO$_2$ electric field energy density at d=200nm & h=5nm

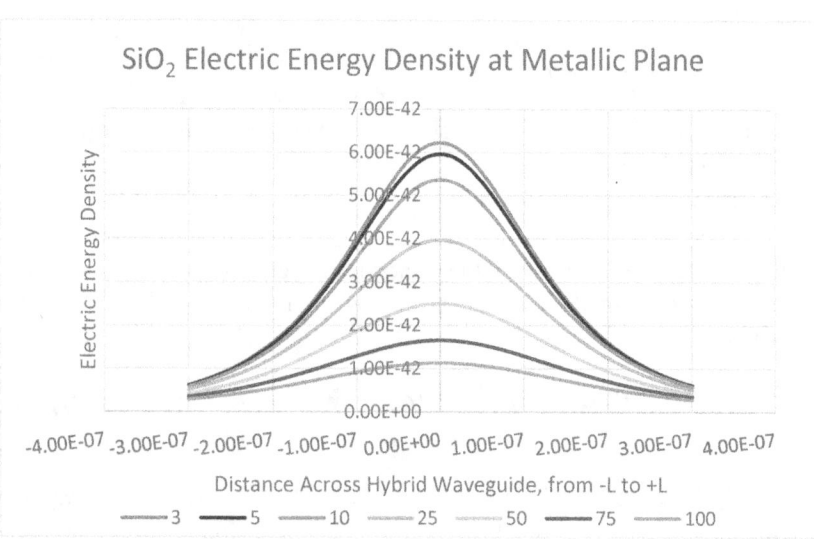

Fig. 4, SiO$_2$ electric field energy density at d=200nm & h=3, 5, 10, 25, 50, 75 &
100nm

A numerical model for (66) & (67) shows (Fig. 2) that E_n is essentially zero, because of the antenna skin effect, this simplifies (65) to,

$$E_n = \delta E \approx 0 \tag{68}$$

$$E_P = E_v - \delta E \tag{69}$$

And from (45) the wrapping efficiency of the nanowire/wave guide structure is,

$$\Theta_W = 10 log \left(\frac{E_P}{E_v}\right) = 10 log \left(1 - \frac{\delta E}{E_v}\right) \tag{70}$$

That is, one requirement for nanowire transmissions is to get as much of the photon energy into the dielectric medium. By (66) this can be effected by reducing the nanowire radius r_n and/or increasing the nanowire dielectric constant ε_n.

Since the charge per length q is a constant, as an example the value[viii] used in this numerical model was *10* electron charges per meter.

Fig 3 & 4, shows how the electric field energy density changes with the height from the metal plane. This model concurs with the Oulton *et al.* [18] proposal that there is capacitor-like energy storage between nanowire and metal plane. However, there is a significant difference between the models. The quantum mechanical model [18] requires a substantial amount of energy within the nanowire while the PSWI nanowire model (66) & (67) treats this as a known electromagnetic skin effect.

Transmission is *wrapping* that occurs along the total length of the nanowire or coaxial cable. From (69) *transmission* is determined by δE or the energy in the nanowire E_n. The larger this value the smaller the energy E_d in the dielectric. From (66) & (57),

$$E_n = \int_0^{r_d} \pi \varepsilon_n \left(P_{\varphi,(x,y,z)}(r/\varepsilon_n 2\pi r_n^2)q\right)^2 r^2 dr \tag{71}$$

$$E_n = \frac{q^2}{4\pi \varepsilon_n r_n^4} \int_0^{r_n} P_{\varphi,(x,y,z)}{}^2 r^4 dr \tag{72}$$

From (36),

$$F_{I,A} = \int_0^{r_n} \pi \varphi_P^2 dr \tag{73}$$

From (3),

$$P_{\varphi,(x,y,z)} = \left(1/k_\psi r\right)/\int_0^{r_n} \pi \varphi_P^2 dr \tag{74}$$

And,

$$E_n = \frac{q^2}{4\pi\varepsilon_n r_n^2} \int_0^{r_n} \left[\frac{1}{k_\psi r \int_0^{r_n} \pi(1/k_\psi r)^2 dr} \right]^2 r^4 dr$$

(75)

$$E_n = \frac{q^2}{\varepsilon_n (2\pi r_n)^2} \int_0^{r_n} \left[\frac{1}{\int_0^{r_n} (1/r)^2 dr} \right]^2 r^2 dr$$

(76)

Replacing the lower limit of the integral of 0 with δ term

$$E_n = \frac{q^2}{\varepsilon_n (2\pi r_n)^2} \int_0^{r_n} \left[\frac{1}{\int_\delta^{r_n} (1/r)^2 dr} \right]^2 r^2 dr$$

(77)

$$E_n = \frac{q^2}{\varepsilon_n (2\pi r_n)^2} \int_0^{r_n} \left[\frac{1}{1/\delta - 1/r_n} \right]^2 r^2 dr$$

(78)

$$E_n = \frac{q^2}{\varepsilon_n (2\pi r_n)^2} \left[\frac{1}{1/\delta - 1/r_n} \right]^2 \int_0^{r_n} r^2 dr$$

(79)

$$E_n = \frac{q^2}{\varepsilon_n (2\pi r_n)^2} \frac{r_n^3}{3} \left[\frac{1}{1/\delta - 1/r_n} \right]^2$$

(80)

$$E_n = K_n \left[\frac{\delta \sqrt{r_n^3}}{r_n - \delta} \right]^2$$

(81)

Where, the *nanowire constant* K_n is given by,

$$K_n = \frac{1}{12\varepsilon_n} \left(\frac{q}{\pi} \right)^2$$

(82)

Therefore in the limit as $\delta \to 0$,

$$E_n = K_n \left[\frac{\delta \sqrt{r_n^3}}{r_n - \delta} \right]^2 \to K_n r_n \delta^2 \to 0$$

(83)

By (69) decreasing the nanowire energy E_n will increase the propagation energy in the dielectric medium E_d. This can be achieved by (81) decreasing the nanowire radius r_n and by increasing its electric permittivity ε_n per (81). The charge per length q, is a function of the photon energy as the photon is the only source of the electric field, and therefore not controllable. However, adding a second *carrier*[ix] photon of a lower frequency could be a method to increasing q and therefore the transmission capability of the nanowire.

By (62) & (63) the third factor that increases the electric field energy E_d in the dielectric medium is the ratio of the dielectrics,

$$\frac{E_{A,d}}{E_{A,n}} = \frac{\varepsilon_n}{\varepsilon_d} \frac{r_n^2}{r^2} \tag{84}$$

That it is not dependent on the charge per length q. Therefore, transmission is primarily dependent upon two nanowire properties r_n, ε_n and the ratio of the electrical permittivities $\varepsilon_n/\varepsilon_d$.

Resolving (34), for a coaxial of outer dielectric radius L,

$$E_d = \int_{r_n}^{L} \pi\varepsilon_d \left(P_{\varphi,(x,y,z)}(1/(\varepsilon_d 2\pi r))q\right)^2 r^2 \, dr \tag{85}$$

$$E_d = \frac{q^2}{4\pi\varepsilon_d} \int_{r_n}^{L} P_{\varphi,(x,y,z)}^2 \, dr \tag{86}$$

From (55),

$$E_d = \frac{q^2}{4\pi^2\varepsilon_d} \int_{r_n}^{L} \left(\varphi_{P,(x,y,z)}/F_{I,A}\right)^2 \, dr \tag{87}$$

From (36),

$$F_{I,A} = \int_{r_n}^{L} \pi\varphi_P^2 \, dr \tag{88}$$

$$E_d = \frac{q^2}{4\pi^4\varepsilon_d} \frac{3k_\psi^2}{[L^3 - r_n^3]^2} \int_{r_n}^{L} \left(\frac{1}{k_\psi r}\right)^2 \, dr \tag{89}$$

$$E_d = \frac{1}{\varepsilon_d}\left(\frac{q}{2\pi^2}\right)^2 \frac{3k_\psi^2}{[L^3 - r_n^3]^2} \frac{[L^3 - r_n^3]}{3k_\psi^2} \tag{90}$$

$$E_d = \frac{1}{\varepsilon_d[L^3 - r_n^3]}\left(\frac{q}{2\pi^2}\right)^2 \tag{91}$$

There are two ways to increase the photon energy E_d in the dielectric medium, i) decrease the dielectric constant ε_d and ii) increase the charge per unit length q.

By (81) & (91) rewriting (46),

$$\frac{E_d}{E_v} = \frac{1}{1 + \left(\frac{\varepsilon_d}{\varepsilon_n}\right)\left(\frac{\delta r_n}{3\pi^2}\right)} \tag{92}$$

That as $\delta \to 0$ or $r_n \to 0$ i.e. the nanowire does not exists, the ratio (92) $\to 1$.

Transmission as opposed to wrapping, has the additional requirement of propagation distance. Propagation distance is reduced by impedance.

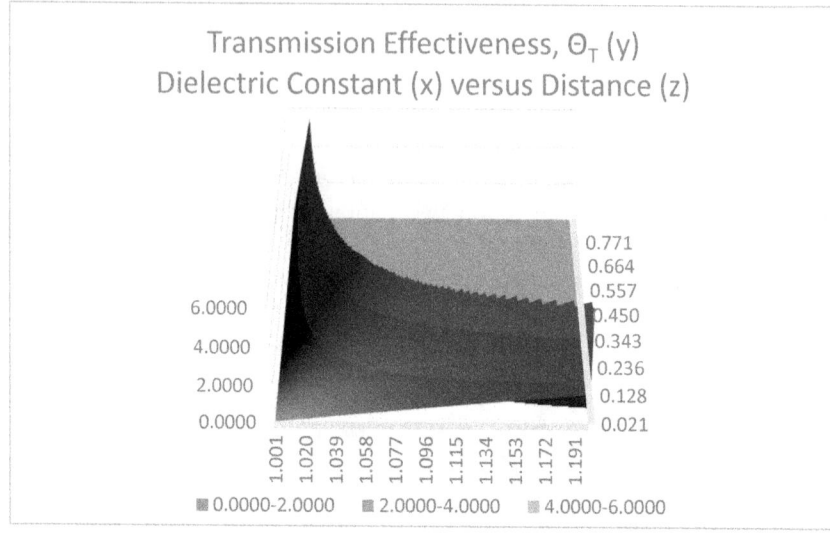

Fig. 5, Transmission Effectiveness Θ_T *as a function of Dielectric Constant and Distance*

Therefore, impedance can be defined as the loss of energy over distance traveled.

Equation (59) defines the maximum electric field strength in the dielectric medium $E_{A,d}$. However, the electric field varies sinusoidally along the z-axis down the nanowire, or

$$_{\alpha}E_{A,d} = \frac{1}{\left(_{\alpha}\varepsilon_d\right)2\pi r} q \sin\frac{2\pi}{\lambda}z \tag{93}$$

$$_{\alpha}E_{A,d} = \frac{1}{\left(_{\alpha}\varepsilon_0\right)2\pi r}\frac{q}{\kappa_d} \sin\frac{2\pi}{\lambda}z \tag{94}$$

Where κ_d is the dielectric constant of the dielectric material.

Since the dielectric material, and matter in general, is not present in the *subspace kenos* β, the electric field strength $_{\alpha}E_{A,d}$ in the *spacetime kenos a* transfers to the *subspace kenos* β as,

$$_{\beta}E_{A,d} = \frac{1}{\varepsilon_0\,2\pi r}\frac{q}{\kappa_d} \sin\frac{2\pi}{\lambda}z \tag{95}$$

This electric field returns to the dielectric material in *spacetime kenos a* a second time,

$$_{\alpha}E_{A,d} = \frac{1}{\left(_{\alpha}\varepsilon_0\right)2\pi r}\frac{q}{\kappa_d^2} \sin\frac{2\pi}{\lambda}2z \tag{96}$$

Therefore, after n cycles, the TEM has reentered the *spacetime kenos a 2n* times (for each half cycle) and propagated a distance $n\lambda$ and (96) is rewritten as,

$$_\alpha E_{A,d} = \frac{1}{\left(_\alpha \varepsilon_0\right) 2\pi r} \frac{q}{\kappa_d^{2n}} \sin \frac{2\pi}{\lambda} nz \tag{97}$$

Therefore, the propagation distance $g\lambda$ is defined by the propagated electric field energy $E_{d,P}$ required at $\zeta = g\lambda$, where g is not an integer or, by (91)

$$E_{d,P} = \frac{1}{\kappa_d^{2g}} \frac{1}{\varepsilon_0 [L^3 - r_n^3]} \left(\frac{q}{2\pi^2}\right)^2 \tag{98}$$

The benchmark energy $E_{d,0}$ in free space is the upper bound of this dielectric electrical energy $E_{d,P}$. Rewriting (98),

$$E_{d,0} = \frac{1}{\varepsilon_0 [L^3 - r_n^3]} \left(\frac{q}{2\pi^2}\right)^2 \tag{99}$$

From (45) the transmission effectiveness Θ_T can be defined as the ratio of this benchmark (99) to its depleted energy (98),

$$\Theta_T = 10 log\left(\frac{E_{d,0}}{E_{d,P}}\right) = 10 log\left(\kappa_d^{2g}\right) \tag{100}$$

The dielectric constant κ_d required to propagate g cycles is,

$$\kappa_d = 10^{(\Theta_T / 20g)} \tag{101}$$

Or for a given dielectric constant κ_d, the propagation distance in g cycles is,

$$g = \frac{\Theta_T}{20 log \kappa_d} \tag{102}$$

Usually the tolerable signal loss per 100ft ($\approx$ 33m) is *5.65dB*. Given that the manufacture of nanowires can produce lengths of have reached *80μm*, substituting *5.65dB* for Θ_T gives,

$$\kappa_d = 1.012588 \tag{103}$$

Or for a SiO_2 dielectric constant of $\kappa_d = 2.25$ gives,

$$g = 0.802141 \tag{104}$$

That is, for a SiO_2 dielectric constant of $\kappa_d = 2.25$ the depth is 0.80λ (or using Oulton *et al* [18] wavelength as an example) the photon does not make it past 1 cycle, or 1,243nm. To be able to propagate the full length of *80μm* would require a dielectric constant κ_d of 1.01 or close to that of vacuum.

For a 14GHz non-air dielectric microwave coaxial cable with the PTFE[x] dielectric $\kappa = 1.3$ gives $g = 2.48$, which means that microwave is reflecting $n = 621.6$.

The plot of (99), Fig. 5[xi], shows that to achieve a reasonable number of reflections, say 10, would require a $\kappa=1.06559$ or close to that of air. The industry term is *effective dielectric constant* which is dependent on transmission line geometries[xii]. An *effective dielectric constant* of $\kappa=1.06559$ suggests a porous dielectric material and reflections due to bends in the coaxial cable.

Therefore, the solutions to increasing nanowire and coaxial cable propagation distances are to engineer porous dielectric materials with dielectric constants close to that of vacuum, implement *carrier* photons, and nanowire radius of at least λ.

7.6 Quantum Tunneling

Returning to quantum tunneling. Its quantum mechanical definition[xiii] is where a electron, is found outside a confining potential despite it having insufficient energy to cross the barrier classically. However, all materials have confining potentials, as materials consists of outer shell valence electrons. Therefore, *tunneling* is the same as *invisibility* but like *wrapping* versus *cloaking*, *tunneling* has the additional requirement that of the charged electron crossing an electrically charged medium. Tunneling effectiveness Θ_U is given by (16),

$$\Theta_U = 10 log \left(\frac{f(P_{\leq r})}{f(P_{\leq R})} \right)$$
$$(105)$$

Given a material's electron electric permittivity $\varepsilon_{e,m}$, the electron's passage through an aperture of radius r_a could be modeled as a short waveguide nanowire, with a nanowire radius of zero. Like the photon's forward motion, the electron has a forward potential motion that is induced by the voltage potential. Another similarity the electron's orthogonal electric field is equivalent to the photon's orthogonal electric field, while electric field along its motion net cancels. This now compares well with the photon's electric field within the waveguide. Thus making available the results derived in subwavelength confinement.

The propagated energy E_P in the aperture/waveguide is the energy between the atoms/molecules E_m. It is the difference between total electric field energy E_v and energy inside the nanowire E_n, and as the nanowire radius is zero, E_n is zero,

$$E_P = E_v$$
$$(106)$$

Rewriting (91) for L= r_a and r_n =0, gives the electron electric field energy E_m passing through the aperture of,

$$E_m = \frac{1}{\varepsilon_{e,m}[r_a^3]}\left(\frac{q}{2\pi^2}\right)^2 \tag{107}$$

That is, the larger the spaces between atoms and molecules the more energy that gets through for a given material's electron electric permittivity $\varepsilon_{e,m}$. For a single aperture, using a hollow vacuum waveguide (no electrical interference from surrounding electron shells) of radius r_a as a bench mark, the total hollow electric energy E_H is,

$$E_H = \frac{1}{\varepsilon_0[r_a^3]}\left(\frac{q}{2\pi^2}\right)^2 \tag{108}$$

gives a tunneling effectiveness Θ_U of,

$$\Theta_U = 10log\left(\frac{E_m}{E_H}\right) = \frac{\varepsilon_0}{\varepsilon_{e,m}} = \frac{1}{\kappa_{e,m}} \tag{109}$$

And for a material with n apertures/waveguides in that material,

$$\Theta_U = 10log\left(\frac{nE_m}{H}\right) = \frac{n\varepsilon_0}{\varepsilon_{e,m}} = \frac{n}{\kappa_{e,m}} \tag{110}$$

Note, that while (109) & (110) are simplified models of electron behavior, just like the photon's electric permittivity ε_m, the materials' electron electric permittivity $\varepsilon_{e,m}$ is a catch all to account for the material's electrical properties, voltage potential, valence electrons, etc.

7.7 Conclusion

The PSWI approach has structured photon behavior into a single phenomenon consisting of three categories shielding, wrapping & invisibility. Thus, clarifying how resolution, cloaking and transmission are variations of wrapping. Since, none of these *effectiveness* equations are direct functions of photon frequency, other than to determine their energy, these phenomena are spectrum independent. Further, the use of the probabilistic *envelope function* φ_P in shielding, wrapping, transmission and invisibility, confirms that particle probabilities are non-Gaussian, and that macro phenomena can be traced back to particle structure.

References:

[1] Solomon, B.T. "Particle Structure", in *Super Physics for Super Technologies*, March 2015.

[2] Solomon, B.T. "Replacing Schrödinger", in *Super Physics for Super Tech-*

nologies, March 2015.

[3] Solomon, B.T. "A Non Standard Model Nucleon/Nuclei Structure", in *Super Physics for Super Technologies*, March 2015.

[4] Solomon, B. T., "The Variable Isotopic Gravitational Constant", in *Super Physics for Super Technologies*, March 2015.

[5] Solomon, B.T., "A Universal Approach to Forces", in *Super Physics for Super Technologies*, March 2015.

[6] Solomon, B. T., "Empirical Evidence Suggest A Need For A Different Gravitational Theory," in the proceedings of the *100 Year Starship Study Public Symposium (100YSS,)*, 2013.

[7] Solomon B.T., "New Evidence, Conditions, Instruments & Experiments for Gravitational Theories", Journal of Modern Physics, Special Issue on Gravitation, Astrophysics and Cosmology, Vol. 8A, 2013, August 2013.

[8] Solomon, B. T., "Empirical Evidence Suggest A Need For A Different Gravitational Theory," *American Physical Society (APS) April Conference, Denver, 2013*

[9] Solomon B.T., *An Introduction to Gravity Modification: A guide to using Laithwaite's and Podkletnov's experiments and the physics of forces for empirical results.* Universal Publishers, Boca Raton, 2nd Edition, May 2012.

[10] Solomon, B. T., "Non-Gaussian Radiation Shielding," *100 Year Starship Study Public Symposium (100YSS,)*, 2011.

[11] Solomon, B.T., "Gravitational Acceleration Without Mass And Noninertia Fields", Physics Essays, Vol. 24, 327, 2011. [Phys. Essays **24**, 327 (2011)]

[12] Solomon, B. T., "Reverse Engineering Podkletnov's Experiments," in the proceedings of the *Space, Propulsion & Energy Sciences International Forum (SPESIF-11)*, Edited by Glen A Robertson, Physics Procedia, Elsevier Science.

[13] Solomon, B. T., "Non-Gaussian Photon Probability Distributions," in the proceedings of the *Space, Propulsion & Energy Sciences International Forum (SPESIF-10)*, Edited by Glen A Robertson, AIP Conference Proceedings **1208**, Melville, New York, (2010).

[14] Solomon, B. T., "An Approach to Gravity Modification as a Propulsion Technology," in the proceedings of the *Space, Propulsion & Energy Sciences International Forum (SPESIF-09)*, Edited by Glen A Robertson, AIP Conference Proceedings **1103**, Melville, New York, (2009).

[15] Solomon, B. T., "An Epiphany On Gravity", Journal of Theoretics, Vol. 3-6, 2001.

[16] Feynman, R., *Feynman Lectures on Physics*, Chapter 28-4.

[17] Nemiroff, R., "Bounds on Spectral Dispersion from Fermi-Detected Gamma Ray Bursts," Physical Review Letters, Vol. 108, No. 23, 2012.

[18] Oulton, R.F., Sorger, V.J., Genov, D.A., Pile, D.F.P. and Zhang, X., "A hybrid plasmonic waveguide for subwavelength confinement and long-range propogation," Nature Photonics, 2, August 2008

[19] Eisaman, M. D., Goldschmidt, E. A., Chen, J., Fan, J. and Migdall, A., "Experimental test of nonlocal realism using a fiber-based source of polarization-entangled photon pairs," *Phys. Rev. A*, **77**, (2008), p. 032339.

[20] Elmore, W. C. and Heald, M.A., *Physics of Waves*, Dover Publications, New York, (1985), p. 241.

[21] Otoshi, T. Y., "A study of microwave transmission through perforated flat plates," JPL Technical Report, 32-1526, Vol. II, (1972).

[22] Prokesa, S. M. and Arnold, S., "Stress-driven formation of Si nanowires", Applied Physics Letters 86, 193105 (2005)

End Notes

i.   Does not include non-peer reviewed conference presentations between 2001 and 2009.
ii. Wrapping is the ability to go around an object as opposed to cloaking which has the additional requirement that the TEM remain undisturbed.
iii. Monotonic relationship, as it allows for the value to be greater if more photon energy gets around the obstruction.
iv. http://hyperphysics.phy-astr.gsu.edu/hbase/electric/engfie.html
v.   http://hyperphysics.phy-astr.gsu.edu/hbase/waves/emwv.html#c1
vi. For radio antennas $\lambda/d>100$.
vii. http://www.allaboutcircuits.com/vol_2/chpt_3/6.html
viii. This value can be changed at any time to fit physical parameters.
ix. The term *carrier* is used because the function of this lower frequency *secondary* photon is to increase the nanowire's ability to *carry* the *primary* higher frequency photon.

x. *ϰ=1.3 is achieved with microporous dielectrics,* http://www.datasheets.pl/coaxial_cables/ PRECISION COAXIAL CABLES OVERVIEW.pdf & http:// www.teledynestorm.com/pdf/HiPITCat_web.pdf
xi. Any *Transmission Effectiveness* Θ_T value greater than 5.65dB is blanked out.
xii. http://www.microwaves101.com/encyclopedias/measuring-dielectric-constant
xiii. http://www.chm.bris.ac.uk/webprojects2000/plewis/Whatis.html